My MATH DRILLS WORKBOOK

TIMED DIVISION WORKSHEETS

REPRODUCIBLE PRACTICE PROBLEMS
(TESTS WITH ANSWERS)

KAROL MORRIS

My Math Drills Workbook – Timed Division Worksheets:
Reproducible Practice Problems (Tests with Answers)

Requests for permission should be sent to:
karol@lastminutemathtutoring.com

ISBN: 978-1-953361-00-4

Table of Contents

NOTE TO TEACHERS, STUDENTS, AND PARENTS V
HOW TO USE THIS WORKBOOK V
MULTIPLICATION AND DIVISION TABLE VI
DIVISION TABLES VII

DIVIDE 0-1 1
DIVIDE 2 5
DIVIDE 0 - 2 9
DIVIDE 3 11
DIVIDE 0 - 3 15
DIVIDE 4 17
DIVIDE 0 - 4 21
DIVIDE 5 23
DIVIDE 0 - 5 27
DIVIDE 6 29
DIVIDE 0 - 6 33
DIVIDE 7 35
DIVIDE 0 - 7 39
DIVIDE 8 41
DIVIDE 0 - 8 45
DIVIDE 9 47
DIVIDE 0 - 9 51
DIVIDE 10 53
DIVIDE 0 - 10 57
DIVIDE 11 59
DIVIDE 0 - 11 63
DIVIDE 12 65
DIVIDE 0 - 12 69
DIVIDE 13 71
DIVIDE 0 - 13 75
ANSWER KEYS 77

ABOUT THE AUTHOR 93

Download Your Free Multiplication Drills!

LINK

NOTE TO TEACHERS, STUDENTS, AND PARENTS

Daily timed math drills provide the repetition needed to memorize math facts and develop math fluency. This workbook contains over 75 days of timed division worksheets, covering 0-13 division facts. These reproducible practice problems may be used for skill development or testing. All answer keys are included in the back.

Teachers and parents may reproduce these timed math tests as needed for their own students and children. This workbook uses an instructional process that the Author has successfully utilized over 25 years as a math instructor. This unique method builds math skills daily while reinforcing math facts learned on prior days.

First, students practice timed division facts involving 0 – 10. Then, division facts of numbers 11 – 13 are introduced. Each number 0 – 13 is practiced four days in a row. To reinforce prior learning, two cumulative division drills are provided after each new number is introduced. Using this repetitive process, students gradually build their division skills by one number at a time while reinforcing earlier division facts mastered.

HOW TO USE THIS WORKBOOK

Set a goal for students to complete one timed division drill per day. Keep track of the time as well as the number of problems answered correctly. Between drills, students may review the division tables provided at the front of the workbook.

Encourage your students to decrease their time while increasing the number of problems answered correctly. Make math fun – set goals and reward students for their progress!

MULTIPLICATION AND DIVISION TABLE

×/÷	0	1	2	3	4	5	6	7	8	9	10	11	12	13
0	0	0	0	0	0	0	0	0	0	0	0	0	0	0
1	0	1	2	3	4	5	6	7	8	9	10	11	12	13
2	0	2	4	6	8	10	12	14	16	18	20	22	24	26
3	0	3	6	9	12	15	18	21	24	27	30	33	36	39
4	0	4	8	12	16	20	24	28	32	36	40	44	48	52
5	0	5	10	15	20	25	30	35	40	45	50	55	60	65
6	0	6	12	18	24	30	36	42	48	54	60	66	72	78
7	0	7	14	21	28	35	42	49	56	63	70	77	84	91
8	0	8	16	24	32	40	48	56	64	72	80	88	96	104
9	0	9	18	27	36	45	54	63	72	81	90	99	108	117
10	0	10	20	30	40	50	60	70	80	90	100	110	120	130
11	0	11	22	33	44	55	66	77	88	99	110	121	132	143
12	0	12	24	36	48	60	72	84	96	108	120	132	144	156
13	0	13	26	39	52	65	78	91	104	117	130	143	156	169

DIVISION TABLES

0 ÷	÷1	÷2	÷3
0 ÷ 1 = 0	1 ÷ 1 = 1	2 ÷ 2 = 1	3 ÷ 3 = 1
0 ÷ 2 = 0	2 ÷ 1 = 2	4 ÷ 2 = 2	6 ÷ 3 = 2
0 ÷ 3 = 0	3 ÷ 1 = 3	6 ÷ 2 = 3	9 ÷ 3 = 3
0 ÷ 4 = 0	4 ÷ 1 = 4	8 ÷ 2 = 4	12 ÷ 3 = 4
0 ÷ 5 = 0	5 ÷ 1 = 5	10 ÷ 2 = 5	15 ÷ 3 = 5
0 ÷ 6 = 0	6 ÷ 1 = 6	12 ÷ 2 = 6	18 ÷ 3 = 6
0 ÷ 7 = 0	7 ÷ 1 = 7	14 ÷ 2 = 7	21 ÷ 3 = 7
0 ÷ 8 = 0	8 ÷ 1 = 8	16 ÷ 2 = 8	24 ÷ 3 = 8
0 ÷ 9 = 0	9 ÷ 1 = 9	18 ÷ 2 = 9	27 ÷ 3 = 9
0 ÷ 10 = 0	10 ÷ 1 = 10	20 ÷ 2 = 10	30 ÷ 3 = 10
0 ÷ 11 = 0	11 ÷ 1 = 11	22 ÷ 2 = 11	33 ÷ 3 = 11
0 ÷ 12 = 0	12 ÷ 1 = 12	24 ÷ 2 = 12	36 ÷ 3 = 12
0 ÷ 13 = 0	13 ÷ 1 = 13	26 ÷ 2 = 13	39 ÷ 3 = 13

÷4	÷5	÷6	÷7
4 ÷ 4 = 1	5 ÷ 5 = 1	6 ÷ 6 = 1	7 ÷ 7 = 1
8 ÷ 4 = 2	10 ÷ 5 = 2	12 ÷ 6 = 2	14 ÷ 7 = 2
12 ÷ 4 = 3	15 ÷ 5 = 3	18 ÷ 6 = 3	21 ÷ 7 = 3
16 ÷ 4 = 4	20 ÷ 5 = 4	24 ÷ 6 = 4	28 ÷ 7 = 4
20 ÷ 4 = 5	25 ÷ 5 = 5	30 ÷ 6 = 5	35 ÷ 7 = 5
24 ÷ 4 = 6	30 ÷ 5 = 6	36 ÷ 6 = 6	42 ÷ 7 = 6
28 ÷ 4 = 7	35 ÷ 5 = 7	42 ÷ 6 = 7	49 ÷ 7 = 7
32 ÷ 4 = 8	40 ÷ 5 = 8	48 ÷ 6 = 8	56 ÷ 7 = 8
36 ÷ 4 = 9	45 ÷ 5 = 9	54 ÷ 6 = 9	63 ÷ 7 = 9
40 ÷ 4 = 10	50 ÷ 5 = 10	60 ÷ 6 = 10	70 ÷ 7 = 10
44 ÷ 4 = 11	55 ÷ 5 = 11	66 ÷ 6 = 11	77 ÷ 7 = 11
48 ÷ 4 = 12	60 ÷ 5 = 12	72 ÷ 6 = 12	84 ÷ 7 = 12
52 ÷ 4 = 13	65 ÷ 5 = 13	78 ÷ 6 = 13	91 ÷ 7 = 13

DIVISION TABLES

÷8	÷9	÷10	÷11
8 ÷ 8 = 1	9 ÷ 9 = 1	10 ÷ 10 = 1	11 ÷ 11 = 1
16 ÷ 8 = 2	18 ÷ 9 = 2	20 ÷ 10 = 2	22 ÷ 11 = 2
24 ÷ 8 = 3	27 ÷ 9 = 3	30 ÷ 10 = 3	33 ÷ 11 = 3
32 ÷ 8 = 4	36 ÷ 9 = 4	40 ÷ 10 = 4	44 ÷ 11 = 4
40 ÷ 8 = 5	45 ÷ 9 = 5	50 ÷ 10 = 5	55 ÷ 11 = 5
48 ÷ 8 = 6	54 ÷ 9 = 6	60 ÷ 10 = 6	66 ÷ 11 = 6
56 ÷ 8 = 7	63 ÷ 9 = 7	70 ÷ 10 = 7	77 ÷ 11 = 7
64 ÷ 8 = 8	72 ÷ 9 = 8	80 ÷ 10 = 8	88 ÷ 11 = 8
72 ÷ 8 = 9	81 ÷ 9 = 9	90 ÷ 10 = 9	99 ÷ 11 = 9
80 ÷ 8 = 10	90 ÷ 9 = 10	100 ÷ 10 = 10	110 ÷ 11 = 10
88 ÷ 8 = 11	99 ÷ 9 = 11	110 ÷ 10 = 11	121 ÷ 11 = 11
96 ÷ 8 = 12	108 ÷ 9 = 12	120 ÷ 10 = 12	132 ÷ 11 = 12
104 ÷ 8 = 13	117 ÷ 9 = 13	130 ÷ 10 = 13	143 ÷ 11 = 13

÷12	÷13
12 ÷ 12 = 1	13 ÷ 13 = 1
24 ÷ 12 = 2	26 ÷ 13 = 2
36 ÷ 12 = 3	39 ÷ 13 = 3
48 ÷ 12 = 4	52 ÷ 13 = 4
60 ÷ 12 = 5	65 ÷ 13 = 5
72 ÷ 12 = 6	78 ÷ 13 = 6
84 ÷ 12 = 7	91 ÷ 13 = 7
96 ÷ 12 = 8	104 ÷ 13 = 8
108 ÷ 12 = 9	117 ÷ 13 = 9
120 ÷ 12 = 10	130 ÷ 13 = 10
132 ÷ 12 = 11	143 ÷ 13 = 11
144 ÷ 12 = 12	156 ÷ 13 = 12
156 ÷ 12 = 13	169 ÷ 13 = 13

This book belongs to
8 ÷ 2 =
21 ÷ 3 =
12 ÷ 4 =

Timed

DIVISION WORKSHEETS

DAY: 1

NAME: ______________________ TIME: ____ : ____ SCORE: ________

DIVIDE 0-1

1)	0 ÷ 5 =	17)	0 ÷ 8 =	33)	0 ÷ 2 =
2)	8 ÷ 1 =	18)	9 ÷ 1 =	34)	4 ÷ 1 =
3)	0 ÷ 3 =	19)	0 ÷ 6 =	35)	0 ÷ 4 =
4)	5 ÷ 1 =	20)	4 ÷ 1 =	36)	3 ÷ 1 =
5)	0 ÷ 2 =	21)	2 ÷ 1 =	37)	0 ÷ 2 =
6)	7 ÷ 1 =	22)	0 ÷ 1 =	38)	3 ÷ 1 =
7)	0 ÷ 6 =	23)	0 ÷ 8 =	39)	0 ÷ 6 =
8)	4 ÷ 1 =	24)	6 ÷ 1 =	40)	0 ÷ 9 =
9)	0 ÷ 8 =	25)	0 ÷ 5 =	41)	4 ÷ 1 =
10)	3 ÷ 1 =	26)	1 ÷ 1 =	42)	1 ÷ 1 =
11)	0 ÷ 7 =	27)	0 ÷ 9 =	43)	0 ÷ 5 =
12)	9 ÷ 1 =	28)	7 ÷ 1 =	44)	5 ÷ 1 =
13)	0 ÷ 1 =	29)	0 ÷ 4 =	45)	0 ÷ 9 =
14)	0 ÷ 6 =	30)	1 ÷ 1 =	46)	6 ÷ 1 =
15)	4 ÷ 1 =	31)	0 ÷ 7 =	47)	0 ÷ 8 =
16)	3 ÷ 1 =	32)	2 ÷ 1 =	48)	8 ÷ 1 =

DAY: 2

NAME: ______________________ TIME: ____ : ____ SCORE: ________

DIVIDE 0-1

1)	0 ÷ 8 =	17)	0 ÷ 3 =	33)	1 ÷ 1 =
2)	9 ÷ 1 =	18)	3 ÷ 1 =	34)	8 ÷ 1 =
3)	4 ÷ 1 =	19)	0 ÷ 7 =	35)	0 ÷ 1 =
4)	7 ÷ 1 =	20)	0 ÷ 4 =	36)	5 ÷ 1 =
5)	0 ÷ 6 =	21)	0 ÷ 1 =	37)	0 ÷ 2 =
6)	3 ÷ 1 =	22)	2 ÷ 1 =	38)	7 ÷ 1 =
7)	0 ÷ 2 =	23)	0 ÷ 8 =	39)	0 ÷ 6 =
8)	4 ÷ 1 =	24)	6 ÷ 1 =	40)	4 ÷ 1 =
9)	3 ÷ 1 =	25)	0 ÷ 5 =	41)	0 ÷ 9 =
10)	0 ÷ 9 =	26)	1 ÷ 1 =	42)	1 ÷ 1 =
11)	0 ÷ 2 =	27)	0 ÷ 7 =	43)	0 ÷ 5 =
12)	3 ÷ 1 =	28)	7 ÷ 1 =	44)	5 ÷ 1 =
13)	0 ÷ 6 =	29)	0 ÷ 4 =	45)	0 ÷ 9 =
14)	4 ÷ 1 =	30)	1 ÷ 1 =	46)	4 ÷ 1 =
15)	9 ÷ 1 =	31)	0 ÷ 9 =	47)	9 ÷ 1 =
16)	0 ÷ 1 =	32)	2 ÷ 1 =	48)	0 ÷ 8 =

DAY: 3

NAME: ____________________ TIME: ____ : ____ SCORE: ________

DIVIDE 0-1

1)	5 ÷ 1 =	17)	6 ÷ 1 =	33)	3 ÷ 1 =
2)	0 ÷ 2 =	18)	0 ÷ 1 =	34)	0 ÷ 2 =
3)	7 ÷ 1 =	19)	2 ÷ 1 =	35)	3 ÷ 1 =
4)	3 ÷ 1 =	20)	8 ÷ 1 =	36)	9 ÷ 1 =
5)	0 ÷ 1 =	21)	0 ÷ 4 =	37)	0 ÷ 9 =
6)	8 ÷ 1 =	22)	9 ÷ 1 =	38)	4 ÷ 1 =
7)	0 ÷ 3 =	23)	0 ÷ 6 =	39)	0 ÷ 9 =
8)	0 ÷ 5 =	24)	0 ÷ 8 =	40)	0 ÷ 2 =
9)	4 ÷ 1 =	25)	1 ÷ 1 =	41)	4 ÷ 1 =
10)	0 ÷ 6 =	26)	0 ÷ 9 =	42)	0 ÷ 8 =
11)	0 ÷ 9 =	27)	0 ÷ 8 =	43)	0 ÷ 6 =
12)	4 ÷ 1 =	28)	6 ÷ 1 =	44)	4 ÷ 1 =
13)	0 ÷ 8 =	29)	0 ÷ 5 =	45)	0 ÷ 9 =
14)	3 ÷ 1 =	30)	1 ÷ 1 =	46)	0 ÷ 4 =
15)	0 ÷ 7 =	31)	0 ÷ 9 =	47)	0 ÷ 5 =
16)	9 ÷ 1 =	32)	7 ÷ 1 =	48)	5 ÷ 1 =

DAY: 4

NAME: ______________________ TIME: ____:____ SCORE: ________

DIVIDE 0-1

1)	3 ÷ 1 =	17)	0 ÷ 8 =	33)	0 ÷ 2 =
2)	0 ÷ 7 =	18)	9 ÷ 1 =	34)	5 ÷ 1 =
3)	9 ÷ 1 =	19)	0 ÷ 6 =	35)	0 ÷ 9 =
4)	0 ÷ 1 =	20)	4 ÷ 1 =	36)	3 ÷ 1 =
5)	4 ÷ 1 =	21)	0 ÷ 1 =	37)	0 ÷ 2 =
6)	0 ÷ 6 =	22)	2 ÷ 1 =	38)	3 ÷ 1 =
7)	3 ÷ 1 =	23)	0 ÷ 8 =	39)	0 ÷ 6 =
8)	4 ÷ 1 =	24)	0 ÷ 5 =	40)	6 ÷ 1 =
9)	0 ÷ 9 =	25)	8 ÷ 1 =	41)	0 ÷ 5 =
10)	1 ÷ 1 =	26)	0 ÷ 3 =	42)	1 ÷ 1 =
11)	0 ÷ 5 =	27)	5 ÷ 1 =	43)	0 ÷ 9 =
12)	0 ÷ 9 =	28)	0 ÷ 2 =	44)	7 ÷ 1 =
13)	5 ÷ 1 =	29)	7 ÷ 1 =	45)	1 ÷ 1 =
14)	4 ÷ 1 =	30)	0 ÷ 6 =	46)	0 ÷ 4 =
15)	0 ÷ 8 =	31)	4 ÷ 1 =	47)	0 ÷ 9 =
16)	9 ÷ 1 =	32)	0 ÷ 8 =	48)	2 ÷ 1 =

DAY: 5

NAME: ____________________ TIME: ___ : ___ SCORE: ________

DIVIDE 2

1)	6 ÷ 2 =	17)	14 ÷ 2 =	33)	10 ÷ 2 =
2)	8 ÷ 2 =	18)	20 ÷ 2 =	34)	8 ÷ 2 =
3)	10 ÷ 2 =	19)	4 ÷ 2 =	35)	10 ÷ 2 =
4)	4 ÷ 2 =	20)	14 ÷ 2 =	36)	16 ÷ 2 =
5)	2 ÷ 2 =	21)	8 ÷ 2 =	37)	6 ÷ 2 =
6)	14 ÷ 2 =	22)	12 ÷ 2 =	38)	4 ÷ 2 =
7)	6 ÷ 2 =	23)	2 ÷ 2 =	39)	12 ÷ 2 =
8)	14 ÷ 2 =	24)	18 ÷ 2 =	40)	8 ÷ 2 =
9)	10 ÷ 2 =	25)	10 ÷ 2 =	41)	6 ÷ 2 =
10)	12 ÷ 2 =	26)	16 ÷ 2 =	42)	2 ÷ 2 =
11)	0 ÷ 2 =	27)	8 ÷ 2 =	43)	4 ÷ 2 =
12)	18 ÷ 2 =	28)	10 ÷ 2 =	44)	12 ÷ 2 =
13)	16 ÷ 2 =	29)	6 ÷ 2 =	45)	4 ÷ 2 =
14)	14 ÷ 2 =	30)	10 ÷ 2 =	46)	20 ÷ 2 =
15)	4 ÷ 2 =	31)	6 ÷ 2 =	47)	18 ÷ 2 =
16)	6 ÷ 2 =	32)	4 ÷ 2 =	48)	8 ÷ 2 =

DAY: 6

NAME: ______________________ TIME: ____ : ____ SCORE: ________

DIVIDE 2

1)	14 ÷ 2 =	17)	6 ÷ 2 =	33)	2 ÷ 2 =
2)	20 ÷ 2 =	18)	8 ÷ 2 =	34)	8 ÷ 2 =
3)	4 ÷ 2 =	19)	10 ÷ 2 =	35)	10 ÷ 2 =
4)	16 ÷ 2 =	20)	4 ÷ 2 =	36)	16 ÷ 2 =
5)	8 ÷ 2 =	21)	2 ÷ 2 =	37)	6 ÷ 2 =
6)	12 ÷ 2 =	22)	14 ÷ 2 =	38)	4 ÷ 2 =
7)	18 ÷ 2 =	23)	6 ÷ 2 =	39)	0 ÷ 2 =
8)	8 ÷ 2 =	24)	4 ÷ 2 =	40)	14 ÷ 2 =
9)	6 ÷ 2 =	25)	10 ÷ 2 =	41)	10 ÷ 2 =
10)	2 ÷ 2 =	26)	16 ÷ 2 =	42)	12 ÷ 2 =
11)	4 ÷ 2 =	27)	8 ÷ 2 =	43)	4 ÷ 2 =
12)	18 ÷ 2 =	28)	12 ÷ 2 =	44)	10 ÷ 2 =
13)	16 ÷ 2 =	29)	4 ÷ 2 =	45)	6 ÷ 2 =
14)	14 ÷ 2 =	30)	20 ÷ 2 =	46)	10 ÷ 2 =
15)	4 ÷ 2 =	31)	18 ÷ 2 =	47)	2 ÷ 2 =
16)	6 ÷ 2 =	32)	8 ÷ 2 =	48)	4 ÷ 2 =

DAY: 7

NAME: ______________________ TIME: ____ : ____ SCORE: ________

DIVIDE 2

1)	8 ÷ 2 =	17)	18 ÷ 2 =	33)	12 ÷ 2 =
2)	14 ÷ 2 =	18)	8 ÷ 2 =	34)	4 ÷ 2 =
3)	10 ÷ 2 =	19)	10 ÷ 2 =	35)	10 ÷ 2 =
4)	12 ÷ 2 =	20)	4 ÷ 2 =	36)	16 ÷ 2 =
5)	4 ÷ 2 =	21)	2 ÷ 2 =	37)	8 ÷ 2 =
6)	10 ÷ 2 =	22)	14 ÷ 2 =	38)	2 ÷ 2 =
7)	12 ÷ 2 =	23)	10 ÷ 2 =	39)	4 ÷ 2 =
8)	8 ÷ 2 =	24)	8 ÷ 2 =	40)	18 ÷ 2 =
9)	6 ÷ 2 =	25)	10 ÷ 2 =	41)	16 ÷ 2 =
10)	12 ÷ 2 =	26)	16 ÷ 2 =	42)	14 ÷ 2 =
11)	14 ÷ 2 =	27)	6 ÷ 2 =	43)	8 ÷ 2 =
12)	20 ÷ 2 =	28)	20 ÷ 2 =	44)	12 ÷ 2 =
13)	4 ÷ 2 =	29)	0 ÷ 2 =	45)	6 ÷ 2 =
14)	14 ÷ 2 =	30)	8 ÷ 2 =	46)	10 ÷ 2 =
15)	2 ÷ 2 =	31)	18 ÷ 2 =	47)	6 ÷ 2 =
16)	6 ÷ 2 =	32)	20 ÷ 2 =	48)	4 ÷ 2 =

DAY: 8

NAME: ______________________ TIME: ____:____ SCORE: ________

DIVIDE 2

1)	8 ÷ 2 =	17)	6 ÷ 2 =	33)	10 ÷ 2 =
2)	14 ÷ 2 =	18)	8 ÷ 2 =	34)	12 ÷ 2 =
3)	10 ÷ 2 =	19)	0 ÷ 2 =	35)	8 ÷ 2 =
4)	12 ÷ 2 =	20)	4 ÷ 2 =	36)	6 ÷ 2 =
5)	4 ÷ 2 =	21)	18 ÷ 2 =	37)	12 ÷ 2 =
6)	12 ÷ 2 =	22)	16 ÷ 2 =	38)	14 ÷ 2 =
7)	4 ÷ 2 =	23)	14 ÷ 2 =	39)	4 ÷ 2 =
8)	10 ÷ 2 =	24)	8 ÷ 2 =	40)	2 ÷ 2 =
9)	16 ÷ 2 =	25)	10 ÷ 2 =	41)	14 ÷ 2 =
10)	8 ÷ 2 =	26)	16 ÷ 2 =	42)	10 ÷ 2 =
11)	2 ÷ 2 =	27)	6 ÷ 2 =	43)	8 ÷ 2 =
12)	20 ÷ 2 =	28)	4 ÷ 2 =	44)	12 ÷ 2 =
13)	4 ÷ 2 =	29)	2 ÷ 2 =	45)	6 ÷ 2 =
14)	14 ÷ 2 =	30)	20 ÷ 2 =	46)	10 ÷ 2 =
15)	16 ÷ 2 =	31)	18 ÷ 2 =	47)	6 ÷ 2 =
16)	6 ÷ 2 =	32)	8 ÷ 2 =	48)	4 ÷ 2 =

DAY: 9

NAME: ______________________ TIME: ____ : ____ SCORE: ________

DIVIDE 0 - 2

1)	6 ÷ 2 =	17)	3 ÷ 1 =	33)	0 ÷ 6 =
2)	5 ÷ 1 =	18)	0 ÷ 2 =	34)	12 ÷ 2 =
3)	4 ÷ 2 =	19)	10 ÷ 2 =	35)	9 ÷ 1 =
4)	12 ÷ 2 =	20)	0 ÷ 1 =	36)	0 ÷ 8 =
5)	7 ÷ 1 =	1)	2 ÷ 1 =	37)	0 ÷ 5 =
6)	0 ÷ 7 =	22)	10 ÷ 2 =	38)	16 ÷ 2 =
7)	20 ÷ 2 =	23)	0 ÷ 9 =	39)	20 ÷ 2 =
8)	6 ÷ 2 =	24)	18 ÷ 2 =	40)	7 ÷ 1 =
9)	2 ÷ 1 =	25)	2 ÷ 1 =	41)	4 ÷ 2 =
10)	2 ÷ 2 =	26)	4 ÷ 1 =	42)	12 ÷ 2 =
11)	0 ÷ 9 =	27)	0 ÷ 3 =	43)	8 ÷ 1 =
12)	8 ÷ 1 =	28)	8 ÷ 2 =	44)	0 ÷ 6 =
13)	0 ÷ 5 =	29)	1 ÷ 1 =	45)	10 ÷ 2 =
14)	18 ÷ 2 =	30)	3 ÷ 1 =	46)	1 ÷ 1 =
15)	0 ÷ 6 =	31)	20 ÷ 2 =	47)	14 ÷ 2 =
16)	6 ÷ 1 =	32)	0 ÷ 4 =	48)	4 ÷ 1 =

DAY: 10

NAME: ______________________ TIME: ____ : ____ SCORE: ________

DIVIDE 0 - 2

1)	8 ÷ 2 =	17)	0 ÷ 6 =	33)	3 ÷ 1 =
2)	6 ÷ 1 =	18)	12 ÷ 2 =	34)	0 ÷ 8 =
3)	4 ÷ 2 =	19)	9 ÷ 1 =	35)	16 ÷ 2 =
4)	0 ÷ 7 =	20)	0 ÷ 2 =	36)	0 ÷ 1 =
5)	9 ÷ 1 =	21)	0 ÷ 5 =	37)	6 ÷ 2 =
6)	8 ÷ 2 =	22)	16 ÷ 2 =	38)	10 ÷ 2 =
7)	20 ÷ 2 =	23)	20 ÷ 2 =	39)	0 ÷ 9 =
8)	0 ÷ 1 =	24)	7 ÷ 1 =	40)	18 ÷ 2 =
9)	2 ÷ 1 =	25)	4 ÷ 2 =	41)	2 ÷ 2 =
10)	0 ÷ 6 =	26)	12 ÷ 2 =	42)	10 ÷ 2 =
11)	0 ÷ 3 =	27)	8 ÷ 1 =	43)	0 ÷ 3 =
12)	5 ÷ 1 =	28)	0 ÷ 6 =	44)	8 ÷ 2 =
13)	14 ÷ 2 =	29)	10 ÷ 2 =	45)	1 ÷ 1 =
14)	18 ÷ 2 =	30)	1 ÷ 1 =	46)	3 ÷ 1 =
15)	8 ÷ 1 =	31)	14 ÷ 2 =	47)	20 ÷ 2 =
16)	6 ÷ 2 =	32)	4 ÷ 1 =	48)	0 ÷ 4 =

DAY: 11

NAME: ______________________ TIME: ____ : ____ SCORE: ________

DIVIDE 3

1)	9 ÷ 3 =	17)	18 ÷ 3 =	33)	3 ÷ 3 =
2)	6 ÷ 3 =	18)	9 ÷ 3 =	34)	18 ÷ 3 =
3)	30 ÷ 3 =	19)	6 ÷ 3 =	35)	15 ÷ 3 =
4)	27 ÷ 3 =	20)	12 ÷ 3 =	36)	12 ÷ 3 =
5)	12 ÷ 3 =	21)	30 ÷ 3 =	37)	18 ÷ 3 =
6)	15 ÷ 3 =	22)	21 ÷ 3 =	38)	9 ÷ 3 =
7)	24 ÷ 3 =	23)	24 ÷ 3 =	39)	12 ÷ 3 =
8)	6 ÷ 3 =	24)	27 ÷ 3 =	40)	6 ÷ 3 =
9)	21 ÷ 3 =	25)	0 ÷ 3 =	41)	9 ÷ 3 =
10)	24 ÷ 3 =	26)	21 ÷ 3 =	42)	12 ÷ 3 =
11)	3 ÷ 3 =	27)	24 ÷ 3 =	43)	3 ÷ 3 =
12)	27 ÷ 3 =	28)	27 ÷ 3 =	44)	15 ÷ 3 =
13)	12 ÷ 3 =	29)	30 ÷ 3 =	45)	18 ÷ 3 =
14)	18 ÷ 3 =	30)	6 ÷ 3 =	46)	30 ÷ 3 =
15)	21 ÷ 3 =	31)	9 ÷ 3 =	47)	3 ÷ 3 =
16)	3 ÷ 3 =	32)	21 ÷ 3 =	48)	18 ÷ 3 =

DAY: 12

NAME: ______________________ TIME: ____ : ____ SCORE: ________

DIVIDE 3

1)	9 ÷ 3 =	17)	3 ÷ 3 =	33)	24 ÷ 3 =
2)	6 ÷ 3 =	18)	18 ÷ 3 =	34)	9 ÷ 3 =
3)	30 ÷ 3 =	19)	15 ÷ 3 =	35)	6 ÷ 3 =
4)	27 ÷ 3 =	20)	12 ÷ 3 =	36)	12 ÷ 3 =
5)	12 ÷ 3 =	21)	18 ÷ 3 =	37)	30 ÷ 3 =
6)	15 ÷ 3 =	22)	0 ÷ 3 =	38)	21 ÷ 3 =
7)	24 ÷ 3 =	23)	12 ÷ 3 =	39)	24 ÷ 3 =
8)	27 ÷ 3 =	24)	6 ÷ 3 =	40)	6 ÷ 3 =
9)	9 ÷ 3 =	25)	9 ÷ 3 =	41)	21 ÷ 3 =
10)	12 ÷ 3 =	26)	21 ÷ 3 =	42)	24 ÷ 3 =
11)	3 ÷ 3 =	27)	24 ÷ 3 =	43)	3 ÷ 3 =
12)	15 ÷ 3 =	28)	27 ÷ 3 =	44)	27 ÷ 3 =
13)	18 ÷ 3 =	29)	30 ÷ 3 =	45)	12 ÷ 3 =
14)	30 ÷ 3 =	30)	6 ÷ 3 =	46)	18 ÷ 3 =
15)	3 ÷ 3 =	31)	9 ÷ 3 =	47)	21 ÷ 3 =
16)	18 ÷ 3 =	32)	21 ÷ 3 =	48)	3 ÷ 3 =

DAY: 13

NAME: ______________________ TIME: ____ : ____ SCORE: ________

DIVIDE 3

1)	3 ÷ 3 =	17)	15 ÷ 3 =	33)	18 ÷ 3 =
2)	27 ÷ 3 =	18)	24 ÷ 3 =	34)	9 ÷ 3 =
3)	12 ÷ 3 =	19)	27 ÷ 3 =	35)	6 ÷ 3 =
4)	18 ÷ 3 =	20)	9 ÷ 3 =	36)	30 ÷ 3 =
5)	21 ÷ 3 =	21)	12 ÷ 3 =	37)	27 ÷ 3 =
6)	0 ÷ 3 =	22)	6 ÷ 3 =	38)	21 ÷ 3 =
7)	18 ÷ 3 =	23)	9 ÷ 3 =	39)	24 ÷ 3 =
8)	15 ÷ 3 =	24)	3 ÷ 3 =	40)	6 ÷ 3 =
9)	12 ÷ 3 =	25)	15 ÷ 3 =	41)	21 ÷ 3 =
10)	18 ÷ 3 =	26)	18 ÷ 3 =	42)	24 ÷ 3 =
11)	9 ÷ 3 =	27)	24 ÷ 3 =	43)	9 ÷ 3 =
12)	12 ÷ 3 =	28)	27 ÷ 3 =	44)	6 ÷ 3 =
13)	21 ÷ 3 =	29)	30 ÷ 3 =	45)	12 ÷ 3 =
14)	30 ÷ 3 =	30)	6 ÷ 3 =	46)	30 ÷ 3 =
15)	24 ÷ 3 =	31)	9 ÷ 3 =	47)	15 ÷ 3 =
16)	18 ÷ 3 =	32)	21 ÷ 3 =	48)	3 ÷ 3 =

DAY: 14

NAME: ______________________ TIME: ___:___ SCORE: ________

DIVIDE 3

1)	15 ÷ 3 =	17)	30 ÷ 3 =	33)	9 ÷ 3 =
2)	9 ÷ 3 =	18)	27 ÷ 3 =	34)	24 ÷ 3 =
3)	3 ÷ 3 =	19)	12 ÷ 3 =	35)	27 ÷ 3 =
4)	30 ÷ 3 =	20)	18 ÷ 3 =	36)	15 ÷ 3 =
5)	27 ÷ 3 =	21)	21 ÷ 3 =	37)	12 ÷ 3 =
6)	21 ÷ 3 =	22)	3 ÷ 3 =	38)	6 ÷ 3 =
7)	24 ÷ 3 =	23)	18 ÷ 3 =	39)	30 ÷ 3 =
8)	6 ÷ 3 =	24)	15 ÷ 3 =	40)	3 ÷ 3 =
9)	21 ÷ 3 =	25)	0 ÷ 3 =	41)	15 ÷ 3 =
10)	24 ÷ 3 =	26)	18 ÷ 3 =	42)	18 ÷ 3 =
11)	9 ÷ 3 =	27)	9 ÷ 3 =	43)	24 ÷ 3 =
12)	6 ÷ 3 =	28)	12 ÷ 3 =	44)	27 ÷ 3 =
13)	12 ÷ 3 =	29)	21 ÷ 3 =	45)	30 ÷ 3 =
14)	30 ÷ 3 =	30)	15 ÷ 3 =	46)	6 ÷ 3 =
15)	12 ÷ 3 =	31)	3 ÷ 3 =	47)	9 ÷ 3 =
16)	0 ÷ 3 =	32)	18 ÷ 3 =	48)	21 ÷ 3 =

DAY: 15

NAME: ____________ TIME: ___ : ___ SCORE: ______

DIVIDE 0 - 3

1)	6 ÷ 3 =	17)	0 ÷ 9 =	33)	1 ÷ 1 =
2)	8 ÷ 2 =	18)	3 ÷ 3 =	34)	15 ÷ 3 =
3)	7 ÷ 1 =	19)	0 ÷ 4 =	35)	9 ÷ 1 =
4)	27 ÷ 3 =	20)	14 ÷ 2 =	36)	5 ÷ 1 =
5)	2 ÷ 1 =	21)	21 ÷ 3 =	37)	6 ÷ 2 =
6)	18 ÷ 3 =	22)	0 ÷ 2 =	38)	6 ÷ 1 =
7)	0 ÷ 1 =	23)	15 ÷ 3 =	39)	12 ÷ 2 =
8)	14 ÷ 2 =	24)	9 ÷ 3 =	40)	0 ÷ 7 =
9)	12 ÷ 3 =	25)	3 ÷ 1 =	41)	18 ÷ 3 =
10)	6 ÷ 1 =	26)	30 ÷ 3 =	42)	2 ÷ 2 =
11)	24 ÷ 3 =	27)	10 ÷ 2 =	43)	24 ÷ 3 =
12)	0 ÷ 6 =	28)	27 ÷ 3 =	44)	12 ÷ 2 =
13)	16 ÷ 2 =	29)	7 ÷ 1 =	45)	15 ÷ 3 =
14)	14 ÷ 2 =	30)	12 ÷ 3 =	46)	20 ÷ 2 =
15)	21 ÷ 3 =	31)	0 ÷ 3 =	47)	18 ÷ 2 =
16)	0 ÷ 8 =	32)	4 ÷ 2 =	48)	0 ÷ 5 =

DAY: 16

NAME: ______________________ TIME: ____ : ____ SCORE: ________

DIVIDE 0 - 3

1)	3 ÷ 1 =	17)	18 ÷ 3 =	33)	12 ÷ 3 =
2)	30 ÷ 3 =	18)	9 ÷ 1 =	34)	1 ÷ 1 =
3)	8 ÷ 2 =	19)	0 ÷ 9 =	35)	24 ÷ 3 =
4)	27 ÷ 3 =	20)	12 ÷ 2 =	36)	0 ÷ 2 =
5)	7 ÷ 1 =	21)	15 ÷ 3 =	37)	16 ÷ 2 =
6)	12 ÷ 3 =	22)	20 ÷ 2 =	38)	14 ÷ 2 =
7)	0 ÷ 6 =	23)	18 ÷ 2 =	39)	21 ÷ 3 =
8)	4 ÷ 2 =	24)	3 ÷ 3 =	40)	0 ÷ 3 =
9)	8 ÷ 1 =	25)	0 ÷ 5 =	41)	6 ÷ 3 =
10)	15 ÷ 3 =	26)	4 ÷ 1 =	42)	8 ÷ 2 =
11)	0 ÷ 4 =	27)	9 ÷ 1 =	43)	7 ÷ 1 =
12)	14 ÷ 2 =	28)	5 ÷ 1 =	44)	27 ÷ 3 =
13)	21 ÷ 3 =	29)	6 ÷ 2 =	45)	2 ÷ 1 =
14)	0 ÷ 8 =	30)	6 ÷ 1 =	46)	18 ÷ 3 =
15)	10 ÷ 2 =	31)	12 ÷ 2 =	47)	0 ÷ 1 =
16)	9 ÷ 3 =	32)	0 ÷ 7 =	48)	2 ÷ 2 =

DAY: 17

NAME: ____________________ TIME: ____ : ____ SCORE: ________

DIVIDE 4

1)	20 ÷ 4 =	17)	40 ÷ 4 =	33)	36 ÷ 4 =
2)	24 ÷ 4 =	18)	16 ÷ 4 =	34)	4 ÷ 4 =
3)	28 ÷ 4 =	19)	8 ÷ 4 =	35)	0 ÷ 4 =
4)	16 ÷ 4 =	20)	12 ÷ 4 =	36)	16 ÷ 4 =
5)	32 ÷ 4 =	21)	32 ÷ 4 =	37)	36 ÷ 4 =
6)	4 ÷ 4 =	22)	8 ÷ 4 =	38)	40 ÷ 4 =
7)	24 ÷ 4 =	23)	16 ÷ 4 =	39)	20 ÷ 4 =
8)	28 ÷ 4 =	24)	12 ÷ 4 =	40)	8 ÷ 4 =
9)	16 ÷ 4 =	25)	0 ÷ 4 =	41)	4 ÷ 4 =
10)	12 ÷ 4 =	26)	8 ÷ 4 =	42)	40 ÷ 4 =
11)	24 ÷ 4 =	27)	20 ÷ 4 =	43)	28 ÷ 4 =
12)	36 ÷ 4 =	28)	8 ÷ 4 =	44)	16 ÷ 4 =
13)	28 ÷ 4 =	29)	12 ÷ 4 =	45)	32 ÷ 4 =
14)	12 ÷ 4 =	30)	36 ÷ 4 =	46)	24 ÷ 4 =
15)	32 ÷ 4 =	31)	40 ÷ 4 =	47)	36 ÷ 4 =
16)	12 ÷ 4 =	32)	20 ÷ 4 =	48)	4 ÷ 4 =

DAY: 18

NAME: ________________ TIME: ____:____ SCORE: ________

DIVIDE 4

1)	40 ÷ 4 =	17)	36 ÷ 4 =	33)	4 ÷ 4 =
2)	12 ÷ 4 =	18)	4 ÷ 4 =	34)	40 ÷ 4 =
3)	24 ÷ 4 =	19)	0 ÷ 4 =	35)	28 ÷ 4 =
4)	36 ÷ 4 =	20)	16 ÷ 4 =	36)	16 ÷ 4 =
5)	28 ÷ 4 =	21)	36 ÷ 4 =	37)	20 ÷ 4 =
6)	12 ÷ 4 =	22)	40 ÷ 4 =	38)	24 ÷ 4 =
7)	32 ÷ 4 =	23)	20 ÷ 4 =	39)	36 ÷ 4 =
8)	20 ÷ 4 =	24)	24 ÷ 4 =	40)	4 ÷ 4 =
9)	12 ÷ 4 =	25)	40 ÷ 4 =	41)	28 ÷ 4 =
10)	24 ÷ 4 =	26)	16 ÷ 4 =	42)	16 ÷ 4 =
11)	36 ÷ 4 =	27)	8 ÷ 4 =	43)	32 ÷ 4 =
12)	8 ÷ 4 =	28)	24 ÷ 4 =	44)	4 ÷ 4 =
13)	20 ÷ 4 =	29)	32 ÷ 4 =	45)	24 ÷ 4 =
14)	8 ÷ 4 =	30)	8 ÷ 4 =	46)	28 ÷ 4 =
15)	12 ÷ 4 =	31)	16 ÷ 4 =	47)	16 ÷ 4 =
16)	40 ÷ 4 =	32)	28 ÷ 4 =	48)	8 ÷ 4 =

DAY: 19

NAME: ______________ TIME: ___:___ SCORE: ______

DIVIDE 4

1)	32 ÷ 4 =	17)	40 ÷ 4 =	33)	8 ÷ 4 =
2)	24 ÷ 4 =	18)	16 ÷ 4 =	34)	16 ÷ 4 =
3)	4 ÷ 4 =	19)	8 ÷ 4 =	35)	12 ÷ 4 =
4)	40 ÷ 4 =	20)	12 ÷ 4 =	36)	36 ÷ 4 =
5)	28 ÷ 4 =	21)	32 ÷ 4 =	37)	4 ÷ 4 =
6)	20 ÷ 4 =	22)	28 ÷ 4 =	38)	12 ÷ 4 =
7)	12 ÷ 4 =	23)	16 ÷ 4 =	39)	16 ÷ 4 =
8)	24 ÷ 4 =	24)	32 ÷ 4 =	40)	8 ÷ 4 =
9)	36 ÷ 4 =	25)	0 ÷ 4 =	41)	4 ÷ 4 =
10)	4 ÷ 4 =	26)	24 ÷ 4 =	42)	40 ÷ 4 =
11)	24 ÷ 4 =	27)	28 ÷ 4 =	43)	20 ÷ 4 =
12)	36 ÷ 4 =	28)	16 ÷ 4 =	44)	8 ÷ 4 =
13)	28 ÷ 4 =	29)	36 ÷ 4 =	45)	0 ÷ 4 =
14)	12 ÷ 4 =	30)	40 ÷ 4 =	46)	36 ÷ 4 =
15)	32 ÷ 4 =	31)	20 ÷ 4 =	47)	40 ÷ 4 =
16)	4 ÷ 4 =	32)	28 ÷ 4 =	48)	24 ÷ 4 =

DAY: 20

NAME: ______________________ TIME: ____ : ____ SCORE: ________

DIVIDE 4

1)	8 ÷ 4 =	17)	32 ÷ 4 =	33)	28 ÷ 4 =
2)	20 ÷ 4 =	18)	12 ÷ 4 =	34)	0 ÷ 4 =
3)	24 ÷ 4 =	19)	28 ÷ 4 =	35)	32 ÷ 4 =
4)	28 ÷ 4 =	20)	20 ÷ 4 =	36)	8 ÷ 4 =
5)	36 ÷ 4 =	21)	16 ÷ 4 =	37)	4 ÷ 4 =
6)	4 ÷ 4 =	22)	32 ÷ 4 =	38)	40 ÷ 4 =
7)	24 ÷ 4 =	23)	4 ÷ 4 =	39)	28 ÷ 4 =
8)	16 ÷ 4 =	24)	8 ÷ 4 =	40)	16 ÷ 4 =
9)	36 ÷ 4 =	25)	12 ÷ 4 =	41)	0 ÷ 4 =
10)	40 ÷ 4 =	26)	36 ÷ 4 =	42)	24 ÷ 4 =
11)	20 ÷ 4 =	27)	40 ÷ 4 =	43)	36 ÷ 4 =
12)	12 ÷ 4 =	28)	20 ÷ 4 =	44)	4 ÷ 4 =
13)	8 ÷ 4 =	29)	40 ÷ 4 =	45)	16 ÷ 4 =
14)	16 ÷ 4 =	30)	16 ÷ 4 =	46)	12 ÷ 4 =
15)	20 ÷ 4 =	31)	8 ÷ 4 =	47)	24 ÷ 4 =
16)	4 ÷ 4 =	32)	12 ÷ 4 =	48)	36 ÷ 4 =

DAY: 21

NAME: ____________________ TIME: ____ : ____ SCORE: ________

DIVIDE 0 - 4

1)	8 ÷ 2 =	1)	3 ÷ 1 =	33)	0 ÷ 8 =
2)	6 ÷ 1 =	2)	0 ÷ 6 =	34)	12 ÷ 4 =
3)	4 ÷ 4 =	3)	16 ÷ 4 =	35)	9 ÷ 1 =
4)	36 ÷ 4 =	4)	3 ÷ 3 =	36)	0 ÷ 5 =
5)	9 ÷ 3 =	5)	21 ÷ 3 =	37)	24 ÷ 4 =
6)	8 ÷ 4 =	6)	10 ÷ 2 =	38)	16 ÷ 2 =
7)	20 ÷ 2 =	7)	0 ÷ 9 =	39)	24 ÷ 3 =
8)	0 ÷ 2 =	8)	18 ÷ 3 =	40)	7 ÷ 1 =
9)	2 ÷ 1 =	9)	14 ÷ 2 =	41)	4 ÷ 2 =
10)	12 ÷ 2 =	10)	40 ÷ 4 =	42)	18 ÷ 2 =
11)	0 ÷ 3 =	11)	27 ÷ 3 =	43)	32 ÷ 4 =
12)	8 ÷ 1 =	12)	6 ÷ 3 =	44)	0 ÷ 1 =
13)	2 ÷ 2 =	13)	30 ÷ 3 =	45)	10 ÷ 2 =
14)	18 ÷ 2 =	14)	28 ÷ 4 =	46)	1 ÷ 1 =
15)	20 ÷ 4 =	15)	20 ÷ 2 =	47)	15 ÷ 3 =
16)	6 ÷ 2 =	16)	0 ÷ 4 =	48)	4 ÷ 1 =

DAY: 22

NAME: ________________ TIME: ___ : ___ SCORE: ________

DIVIDE 0 - 4

1)	0 ÷ 8 =	17)	8 ÷ 2 =	33)	3 ÷ 1 =
2)	40 ÷ 4 =	18)	6 ÷ 1 =	34)	0 ÷ 1 =
3)	9 ÷ 1 =	19)	4 ÷ 4 =	35)	16 ÷ 4 =
4)	0 ÷ 5 =	20)	2 ÷ 2 =	36)	9 ÷ 3 =
5)	24 ÷ 4 =	21)	9 ÷ 3 =	37)	24 ÷ 4 =
6)	16 ÷ 2 =	22)	8 ÷ 4 =	38)	10 ÷ 2 =
7)	21 ÷ 3 =	23)	20 ÷ 2 =	39)	0 ÷ 9 =
8)	32 ÷ 4 =	24)	0 ÷ 2 =	40)	18 ÷ 3 =
9)	4 ÷ 2 =	25)	2 ÷ 1 =	41)	14 ÷ 2 =
10)	18 ÷ 2 =	26)	12 ÷ 2 =	42)	12 ÷ 4 =
11)	8 ÷ 4 =	27)	0 ÷ 3 =	43)	27 ÷ 3 =
12)	0 ÷ 6 =	28)	8 ÷ 1 =	44)	21 ÷ 3 =
13)	10 ÷ 2 =	29)	24 ÷ 3 =	45)	7 ÷ 1 =
14)	1 ÷ 1 =	30)	18 ÷ 2 =	46)	28 ÷ 4 =
15)	15 ÷ 3 =	31)	20 ÷ 4 =	47)	30 ÷ 3 =
16)	4 ÷ 1 =	32)	6 ÷ 2 =	48)	0 ÷ 4 =

DAY: 23

NAME: ______________________ TIME: ____ : ____ SCORE: ________

DIVIDE 5

1)	35 ÷ 5 =	17)	40 ÷ 5 =	33)	15 ÷ 5 =
2)	10 ÷ 5 =	18)	15 ÷ 5 =	34)	10 ÷ 5 =
3)	15 ÷ 5 =	19)	50 ÷ 5 =	35)	5 ÷ 5 =
4)	40 ÷ 5 =	20)	10 ÷ 5 =	36)	30 ÷ 5 =
5)	30 ÷ 5 =	21)	30 ÷ 5 =	37)	25 ÷ 5 =
6)	20 ÷ 5 =	22)	35 ÷ 5 =	38)	45 ÷ 5 =
7)	25 ÷ 5 =	23)	20 ÷ 5 =	39)	30 ÷ 5 =
8)	45 ÷ 5 =	24)	35 ÷ 5 =	40)	5 ÷ 5 =
9)	5 ÷ 5 =	25)	15 ÷ 5 =	41)	50 ÷ 5 =
10)	15 ÷ 5 =	26)	25 ÷ 5 =	42)	40 ÷ 5 =
11)	20 ÷ 5 =	27)	40 ÷ 5 =	43)	35 ÷ 5 =
12)	10 ÷ 5 =	28)	0 ÷ 5 =	44)	50 ÷ 5 =
13)	50 ÷ 5 =	29)	10 ÷ 5 =	45)	25 ÷ 5 =
14)	40 ÷ 5 =	30)	30 ÷ 5 =	46)	5 ÷ 5 =
15)	20 ÷ 5 =	31)	40 ÷ 5 =	47)	45 ÷ 5 =
16)	45 ÷ 5 =	32)	20 ÷ 5 =	48)	40 ÷ 5 =

DAY: 24

NAME: ______________________ TIME: ___:___ SCORE: ________

DIVIDE 5

1)	15 ÷ 5 =	17)	40 ÷ 5 =	33)	35 ÷ 5 =
2)	10 ÷ 5 =	18)	45 ÷ 5 =	34)	10 ÷ 5 =
3)	5 ÷ 5 =	19)	50 ÷ 5 =	35)	15 ÷ 5 =
4)	30 ÷ 5 =	20)	10 ÷ 5 =	36)	40 ÷ 5 =
5)	25 ÷ 5 =	21)	0 ÷ 5 =	37)	30 ÷ 5 =
6)	45 ÷ 5 =	22)	35 ÷ 5 =	38)	20 ÷ 5 =
7)	25 ÷ 5 =	23)	20 ÷ 5 =	39)	30 ÷ 5 =
8)	0 ÷ 5 =	24)	45 ÷ 5 =	40)	5 ÷ 5 =
9)	5 ÷ 5 =	25)	50 ÷ 5 =	41)	15 ÷ 5 =
10)	15 ÷ 5 =	26)	40 ÷ 5 =	42)	25 ÷ 5 =
11)	20 ÷ 5 =	27)	35 ÷ 5 =	43)	40 ÷ 5 =
12)	10 ÷ 5 =	28)	50 ÷ 5 =	44)	5 ÷ 5 =
13)	50 ÷ 5 =	29)	10 ÷ 5 =	45)	35 ÷ 5 =
14)	40 ÷ 5 =	30)	5 ÷ 5 =	46)	30 ÷ 5 =
15)	20 ÷ 5 =	31)	50 ÷ 5 =	47)	45 ÷ 5 =
16)	45 ÷ 5 =	32)	20 ÷ 5 =	48)	40 ÷ 5 =

DAY: 25

NAME: ______________________ TIME: ____:____ SCORE: ________

DIVIDE 5

1)	15 ÷ 5 =	17)	45 ÷ 5 =	33)	35 ÷ 5 =
2)	20 ÷ 5 =	18)	15 ÷ 5 =	34)	10 ÷ 5 =
3)	35 ÷ 5 =	19)	50 ÷ 5 =	35)	15 ÷ 5 =
4)	50 ÷ 5 =	20)	25 ÷ 5 =	36)	40 ÷ 5 =
5)	25 ÷ 5 =	21)	40 ÷ 5 =	37)	30 ÷ 5 =
6)	45 ÷ 5 =	22)	5 ÷ 5 =	38)	20 ÷ 5 =
7)	25 ÷ 5 =	23)	10 ÷ 5 =	39)	35 ÷ 5 =
8)	45 ÷ 5 =	24)	30 ÷ 5 =	40)	5 ÷ 5 =
9)	5 ÷ 5 =	25)	45 ÷ 5 =	41)	15 ÷ 5 =
10)	15 ÷ 5 =	26)	40 ÷ 5 =	42)	10 ÷ 5 =
11)	10 ÷ 5 =	27)	35 ÷ 5 =	43)	30 ÷ 5 =
12)	5 ÷ 5 =	28)	50 ÷ 5 =	44)	35 ÷ 5 =
13)	30 ÷ 5 =	29)	10 ÷ 5 =	45)	25 ÷ 5 =
14)	40 ÷ 5 =	30)	0 ÷ 5 =	46)	45 ÷ 5 =
15)	20 ÷ 5 =	31)	40 ÷ 5 =	47)	5 ÷ 5 =
16)	45 ÷ 5 =	32)	20 ÷ 5 =	48)	50 ÷ 5 =

DAY: 26

NAME: ____________________ TIME: ____ : ____ SCORE: ________

DIVIDE 5

1)	30 ÷ 5 =	17)	45 ÷ 5 =	33)	35 ÷ 5 =
2)	5 ÷ 5 =	18)	0 ÷ 5 =	34)	10 ÷ 5 =
3)	15 ÷ 5 =	19)	15 ÷ 5 =	35)	15 ÷ 5 =
4)	10 ÷ 5 =	20)	35 ÷ 5 =	36)	40 ÷ 5 =
5)	30 ÷ 5 =	21)	5 ÷ 5 =	37)	30 ÷ 5 =
6)	45 ÷ 5 =	22)	45 ÷ 5 =	38)	20 ÷ 5 =
7)	5 ÷ 5 =	23)	10 ÷ 5 =	39)	15 ÷ 5 =
8)	40 ÷ 5 =	24)	30 ÷ 5 =	40)	20 ÷ 5 =
9)	15 ÷ 5 =	25)	45 ÷ 5 =	41)	10 ÷ 5 =
10)	50 ÷ 5 =	26)	40 ÷ 5 =	42)	50 ÷ 5 =
11)	25 ÷ 5 =	27)	35 ÷ 5 =	43)	25 ÷ 5 =
12)	40 ÷ 5 =	28)	50 ÷ 5 =	44)	35 ÷ 5 =
13)	30 ÷ 5 =	29)	10 ÷ 5 =	45)	20 ÷ 5 =
14)	40 ÷ 5 =	30)	5 ÷ 5 =	46)	45 ÷ 5 =
15)	20 ÷ 5 =	31)	25 ÷ 5 =	47)	50 ÷ 5 =
16)	45 ÷ 5 =	32)	20 ÷ 5 =	48)	40 ÷ 5 =

DAY: 27

NAME: ________________ TIME: ____:____ SCORE: ________

DIVIDE 0 - 5

1)	20 ÷ 4 =	17)	2 ÷ 2 =	33)	10 ÷ 2 =
2)	5 ÷ 5 =	18)	15 ÷ 5 =	34)	12 ÷ 4 =
3)	35 ÷ 5 =	19)	32 ÷ 4 =	35)	12 ÷ 2 =
4)	27 ÷ 3 =	20)	12 ÷ 3 =	36)	0 ÷ 4 =
5)	6 ÷ 2 =	21)	0 ÷ 3 =	37)	36 ÷ 4 =
6)	3 ÷ 3 =	22)	8 ÷ 4 =	38)	45 ÷ 5 =
7)	9 ÷ 1 =	23)	16 ÷ 2 =	39)	0 ÷ 5 =
8)	14 ÷ 2 =	24)	20 ÷ 2 =	40)	28 ÷ 4 =
9)	16 ÷ 4 =	25)	24 ÷ 4 =	41)	18 ÷ 3 =
10)	30 ÷ 5 =	26)	18 ÷ 2 =	42)	1 ÷ 1 =
11)	15 ÷ 3 =	27)	4 ÷ 4 =	43)	24 ÷ 3 =
12)	0 ÷ 9 =	28)	25 ÷ 5 =	44)	8 ÷ 2 =
13)	50 ÷ 5 =	29)	9 ÷ 3 =	45)	15 ÷ 3 =
14)	0 ÷ 10 =	30)	4 ÷ 2 =	46)	8 ÷ 1 =
15)	21 ÷ 3 =	31)	10 ÷ 1 =	47)	40 ÷ 5 =
16)	0 ÷ 1 =	32)	40 ÷ 4 =	48)	30 ÷ 3 =

DAY: 28

NAME: ____________________ TIME: ___ : ___ SCORE: ________

DIVIDE 0 - 5

1)	2 ÷ 2 =	17)	10 ÷ 2 =	33)	20 ÷ 4 =
2)	15 ÷ 5 =	18)	12 ÷ 4 =	34)	24 ÷ 3 =
3)	32 ÷ 4 =	19)	9 ÷ 1 =	35)	35 ÷ 5 =
4)	12 ÷ 3 =	20)	0 ÷ 1 =	36)	27 ÷ 3 =
5)	0 ÷ 3 =	21)	36 ÷ 4 =	37)	6 ÷ 2 =
6)	8 ÷ 4 =	22)	45 ÷ 5 =	38)	6 ÷ 3 =
7)	16 ÷ 2 =	23)	0 ÷ 5 =	39)	9 ÷ 1 =
8)	9 ÷ 3 =	24)	28 ÷ 4 =	40)	14 ÷ 2 =
9)	24 ÷ 4 =	25)	18 ÷ 3 =	41)	16 ÷ 4 =
10)	18 ÷ 2 =	26)	50 ÷ 5 =	42)	30 ÷ 5 =
11)	4 ÷ 4 =	27)	24 ÷ 4 =	43)	15 ÷ 3 =
12)	25 ÷ 5 =	28)	8 ÷ 2 =	44)	0 ÷ 9 =
13)	3 ÷ 3 =	29)	15 ÷ 3 =	45)	20 ÷ 2 =
14)	4 ÷ 2 =	30)	8 ÷ 1 =	46)	0 ÷ 4 =
15)	10 ÷ 1 =	31)	40 ÷ 5 =	47)	21 ÷ 3 =
16)	12 ÷ 2 =	32)	40 ÷ 4 =	48)	0 ÷ 2 =

DAY: 29

NAME: ______________________ TIME: ____:____ SCORE: ________

DIVIDE 6

1)	36 ÷ 6 =	17)	42 ÷ 6 =	33)	30 ÷ 6 =
2)	6 ÷ 6 =	18)	18 ÷ 6 =	34)	12 ÷ 6 =
3)	18 ÷ 6 =	19)	60 ÷ 6 =	35)	54 ÷ 6 =
4)	12 ÷ 6 =	20)	0 ÷ 6 =	36)	60 ÷ 6 =
5)	30 ÷ 6 =	21)	30 ÷ 6 =	37)	24 ÷ 6 =
6)	24 ÷ 6 =	22)	36 ÷ 6 =	38)	36 ÷ 6 =
7)	54 ÷ 6 =	23)	48 ÷ 6 =	39)	30 ÷ 6 =
8)	42 ÷ 6 =	24)	42 ÷ 6 =	40)	12 ÷ 6 =
9)	48 ÷ 6 =	25)	54 ÷ 6 =	41)	60 ÷ 6 =
10)	6 ÷ 6 =	26)	12 ÷ 6 =	42)	48 ÷ 6 =
11)	12 ÷ 6 =	27)	18 ÷ 6 =	43)	36 ÷ 6 =
12)	18 ÷ 6 =	28)	6 ÷ 6 =	44)	24 ÷ 6 =
13)	60 ÷ 6 =	29)	54 ÷ 6 =	45)	0 ÷ 6 =
14)	36 ÷ 6 =	30)	42 ÷ 6 =	46)	6 ÷ 6 =
15)	48 ÷ 6 =	31)	24 ÷ 6 =	47)	30 ÷ 6 =
16)	60 ÷ 6 =	32)	36 ÷ 6 =	48)	42 ÷ 6 =

DAY: 30

NAME: ____________________ TIME: ___:___ SCORE: ________

DIVIDE 6

1)	30 ÷ 6 =	17)	42 ÷ 6 =	33)	24 ÷ 6 =
2)	12 ÷ 6 =	18)	18 ÷ 6 =	34)	36 ÷ 6 =
3)	54 ÷ 6 =	19)	48 ÷ 6 =	35)	60 ÷ 6 =
4)	60 ÷ 6 =	20)	42 ÷ 6 =	36)	6 ÷ 6 =
5)	36 ÷ 6 =	21)	54 ÷ 6 =	37)	30 ÷ 6 =
6)	6 ÷ 6 =	22)	0 ÷ 6 =	38)	36 ÷ 6 =
7)	18 ÷ 6 =	23)	30 ÷ 6 =	39)	54 ÷ 6 =
8)	12 ÷ 6 =	24)	12 ÷ 6 =	40)	42 ÷ 6 =
9)	30 ÷ 6 =	25)	60 ÷ 6 =	41)	48 ÷ 6 =
10)	24 ÷ 6 =	26)	48 ÷ 6 =	42)	24 ÷ 6 =
11)	18 ÷ 6 =	27)	12 ÷ 6 =	43)	36 ÷ 6 =
12)	0 ÷ 6 =	28)	6 ÷ 6 =	44)	60 ÷ 6 =
13)	60 ÷ 6 =	29)	54 ÷ 6 =	45)	18 ÷ 6 =
14)	36 ÷ 6 =	30)	42 ÷ 6 =	46)	6 ÷ 6 =
15)	48 ÷ 6 =	31)	48 ÷ 6 =	47)	30 ÷ 6 =
16)	24 ÷ 6 =	32)	36 ÷ 6 =	48)	42 ÷ 6 =

DAY: 31

NAME: ____________ TIME: ___:___ SCORE: ______

DIVIDE 6

1)	30 ÷ 6 =	17)	42 ÷ 6 =	33)	36 ÷ 6 =
2)	24 ÷ 6 =	18)	18 ÷ 6 =	34)	6 ÷ 6 =
3)	54 ÷ 6 =	19)	48 ÷ 6 =	35)	18 ÷ 6 =
4)	60 ÷ 6 =	20)	6 ÷ 6 =	36)	12 ÷ 6 =
5)	24 ÷ 6 =	21)	30 ÷ 6 =	37)	30 ÷ 6 =
6)	0 ÷ 6 =	22)	36 ÷ 6 =	38)	24 ÷ 6 =
7)	48 ÷ 6 =	23)	54 ÷ 6 =	39)	30 ÷ 6 =
8)	42 ÷ 6 =	24)	0 ÷ 6 =	40)	12 ÷ 6 =
9)	54 ÷ 6 =	25)	48 ÷ 6 =	41)	60 ÷ 6 =
10)	12 ÷ 6 =	26)	6 ÷ 6 =	42)	48 ÷ 6 =
11)	18 ÷ 6 =	27)	12 ÷ 6 =	43)	36 ÷ 6 =
12)	24 ÷ 6 =	28)	6 ÷ 6 =	44)	60 ÷ 6 =
13)	60 ÷ 6 =	29)	54 ÷ 6 =	45)	18 ÷ 6 =
14)	36 ÷ 6 =	30)	42 ÷ 6 =	46)	6 ÷ 6 =
15)	48 ÷ 6 =	31)	0 ÷ 6 =	47)	30 ÷ 6 =
16)	60 ÷ 6 =	32)	36 ÷ 6 =	48)	42 ÷ 6 =

DAY: 32

NAME: ____________ TIME: ___ : ___ SCORE: ______

DIVIDE 6

1)	42 ÷ 6 =	17)	42 ÷ 6 =	33)	24 ÷ 6 =
2)	48 ÷ 6 =	18)	18 ÷ 6 =	34)	36 ÷ 6 =
3)	6 ÷ 6 =	19)	48 ÷ 6 =	35)	60 ÷ 6 =
4)	36 ÷ 6 =	20)	42 ÷ 6 =	36)	6 ÷ 6 =
5)	12 ÷ 6 =	21)	54 ÷ 6 =	37)	30 ÷ 6 =
6)	6 ÷ 6 =	22)	12 ÷ 6 =	38)	36 ÷ 6 =
7)	18 ÷ 6 =	23)	30 ÷ 6 =	39)	54 ÷ 6 =
8)	12 ÷ 6 =	24)	0 ÷ 6 =	40)	30 ÷ 6 =
9)	30 ÷ 6 =	25)	60 ÷ 6 =	41)	12 ÷ 6 =
10)	24 ÷ 6 =	26)	48 ÷ 6 =	42)	54 ÷ 6 =
11)	6 ÷ 6 =	27)	12 ÷ 6 =	43)	60 ÷ 6 =
12)	18 ÷ 6 =	28)	24 ÷ 6 =	44)	6 ÷ 6 =
13)	60 ÷ 6 =	29)	54 ÷ 6 =	45)	12 ÷ 6 =
14)	36 ÷ 6 =	30)	42 ÷ 6 =	46)	36 ÷ 6 =
15)	48 ÷ 6 =	31)	48 ÷ 6 =	47)	30 ÷ 6 =
16)	60 ÷ 6 =	32)	36 ÷ 6 =	48)	42 ÷ 6 =

DAY: 33

NAME: ____________ TIME: ___ : ___ SCORE: ______

DIVIDE 0 - 6

1)	24 ÷ 6 =	17)	2 ÷ 1 =	33)	10 ÷ 5 =
2)	18 ÷ 2 =	18)	16 ÷ 2 =	34)	42 ÷ 6 =
3)	35 ÷ 5 =	19)	32 ÷ 4 =	35)	5 ÷ 1 =
4)	28 ÷ 4 =	20)	27 ÷ 3 =	36)	36 ÷ 6 =
5)	6 ÷ 3 =	21)	0 ÷ 7 =	37)	24 ÷ 4 =
6)	0 ÷ 9 =	22)	36 ÷ 4 =	38)	45 ÷ 5 =
7)	9 ÷ 3 =	23)	54 ÷ 6 =	39)	0 ÷ 5 =
8)	12 ÷ 2 =	24)	35 ÷ 5 =	40)	6 ÷ 6 =
9)	16 ÷ 4 =	25)	20 ÷ 4 =	41)	14 ÷ 2 =
10)	30 ÷ 6 =	26)	18 ÷ 6 =	42)	6 ÷ 1 =
11)	15 ÷ 5 =	27)	4 ÷ 1 =	43)	12 ÷ 6 =
12)	0 ÷ 2 =	28)	25 ÷ 5 =	44)	6 ÷ 2 =
13)	16 ÷ 2 =	29)	9 ÷ 3 =	45)	21 ÷ 3 =
14)	1 ÷ 1 =	30)	8 ÷ 2 =	46)	8 ÷ 1 =
15)	15 ÷ 3 =	31)	9 ÷ 1 =	47)	60 ÷ 6 =
16)	0 ÷ 1 =	32)	40 ÷ 4 =	48)	40 ÷ 5 =

DAY: 34

NAME: ____________ TIME: ___:___ SCORE: ________

DIVIDE 0 - 6

1)	2 ÷ 1 =	17)	10 ÷ 5 =	33)	24 ÷ 6 =
2)	16 ÷ 2 =	18)	12 ÷ 2 =	34)	18 ÷ 2 =
3)	32 ÷ 4 =	19)	5 ÷ 1 =	35)	35 ÷ 5 =
4)	27 ÷ 3 =	20)	36 ÷ 6 =	36)	28 ÷ 4 =
5)	0 ÷ 7 =	21)	24 ÷ 4 =	37)	6 ÷ 3 =
6)	36 ÷ 4 =	22)	45 ÷ 5 =	38)	0 ÷ 9 =
7)	54 ÷ 6 =	23)	0 ÷ 5 =	39)	18 ÷ 3 =
8)	35 ÷ 5 =	24)	6 ÷ 6 =	40)	12 ÷ 3 =
9)	20 ÷ 4 =	25)	14 ÷ 2 =	41)	16 ÷ 4 =
10)	18 ÷ 6 =	26)	6 ÷ 1 =	42)	30 ÷ 6 =
11)	4 ÷ 1 =	27)	12 ÷ 6 =	43)	15 ÷ 5 =
12)	25 ÷ 5 =	28)	6 ÷ 2 =	44)	0 ÷ 2 =
13)	9 ÷ 3 =	29)	21 ÷ 3 =	45)	20 ÷ 2 =
14)	8 ÷ 2 =	30)	8 ÷ 1 =	46)	1 ÷ 1 =
15)	9 ÷ 1 =	31)	60 ÷ 6 =	47)	15 ÷ 3 =
16)	40 ÷ 4 =	32)	40 ÷ 5 =	48)	0 ÷ 1 =

DAY: 35

NAME: ______________________ TIME: ____:____ SCORE: ________

DIVIDE 7

1)	21 ÷ 7 =	17)	42 ÷ 7 =	33)	35 ÷ 7 =
2)	7 ÷ 7 =	18)	14 ÷ 7 =	34)	28 ÷ 7 =
3)	0 ÷ 7 =	19)	49 ÷ 7 =	35)	56 ÷ 7 =
4)	28 ÷ 7 =	20)	7 ÷ 7 =	36)	70 ÷ 7 =
5)	35 ÷ 7 =	21)	56 ÷ 7 =	37)	63 ÷ 7 =
6)	49 ÷ 7 =	22)	28 ÷ 7 =	38)	42 ÷ 7 =
7)	21 ÷ 7 =	23)	56 ÷ 7 =	39)	14 ÷ 7 =
8)	42 ÷ 7 =	24)	63 ÷ 7 =	40)	21 ÷ 7 =
9)	56 ÷ 7 =	25)	70 ÷ 7 =	41)	70 ÷ 7 =
10)	28 ÷ 7 =	26)	0 ÷ 7 =	42)	49 ÷ 7 =
11)	70 ÷ 7 =	27)	14 ÷ 7 =	43)	42 ÷ 7 =
12)	63 ÷ 7 =	28)	28 ÷ 7 =	44)	7 ÷ 7 =
13)	21 ÷ 7 =	29)	35 ÷ 7 =	45)	14 ÷ 7 =
14)	35 ÷ 7 =	30)	21 ÷ 7 =	46)	70 ÷ 7 =
15)	49 ÷ 7 =	31)	63 ÷ 7 =	47)	35 ÷ 7 =
16)	7 ÷ 7 =	32)	14 ÷ 7 =	48)	42 ÷ 7 =

DAY: 36

NAME: ______________________ TIME: ____:____ SCORE: ________

DIVIDE 7

1)	42 ÷ 7 =	17)	21 ÷ 7 =	33)	35 ÷ 7 =
2)	14 ÷ 7 =	18)	7 ÷ 7 =	34)	28 ÷ 7 =
3)	49 ÷ 7 =	19)	14 ÷ 7 =	35)	56 ÷ 7 =
4)	56 ÷ 7 =	20)	28 ÷ 7 =	36)	70 ÷ 7 =
5)	28 ÷ 7 =	21)	35 ÷ 7 =	37)	63 ÷ 7 =
6)	14 ÷ 7 =	22)	49 ÷ 7 =	38)	42 ÷ 7 =
7)	28 ÷ 7 =	23)	0 ÷ 7 =	39)	14 ÷ 7 =
8)	63 ÷ 7 =	24)	49 ÷ 7 =	40)	21 ÷ 7 =
9)	70 ÷ 7 =	25)	7 ÷ 7 =	41)	70 ÷ 7 =
10)	7 ÷ 7 =	26)	56 ÷ 7 =	42)	49 ÷ 7 =
11)	70 ÷ 7 =	27)	28 ÷ 7 =	43)	42 ÷ 7 =
12)	63 ÷ 7 =	28)	56 ÷ 7 =	44)	7 ÷ 7 =
13)	21 ÷ 7 =	29)	35 ÷ 7 =	45)	0 ÷ 7 =
14)	35 ÷ 7 =	30)	21 ÷ 7 =	46)	70 ÷ 7 =
15)	49 ÷ 7 =	31)	63 ÷ 7 =	47)	35 ÷ 7 =
16)	7 ÷ 7 =	32)	14 ÷ 7 =	48)	42 ÷ 7 =

DAY: 37

NAME: ______________________ TIME: ____ : ____ SCORE: ________

DIVIDE 7

1)	56 ÷ 7 =	17)	21 ÷ 7 =	33)	35 ÷ 7 =
2)	70 ÷ 7 =	18)	7 ÷ 7 =	34)	28 ÷ 7 =
3)	63 ÷ 7 =	19)	14 ÷ 7 =	35)	42 ÷ 7 =
4)	35 ÷ 7 =	20)	28 ÷ 7 =	36)	56 ÷ 7 =
5)	14 ÷ 7 =	21)	35 ÷ 7 =	37)	42 ÷ 7 =
6)	56 ÷ 7 =	22)	49 ÷ 7 =	38)	14 ÷ 7 =
7)	28 ÷ 7 =	23)	14 ÷ 7 =	39)	28 ÷ 7 =
8)	21 ÷ 7 =	24)	49 ÷ 7 =	40)	63 ÷ 7 =
9)	70 ÷ 7 =	25)	7 ÷ 7 =	41)	70 ÷ 7 =
10)	7 ÷ 7 =	26)	56 ÷ 7 =	42)	49 ÷ 7 =
11)	70 ÷ 7 =	27)	28 ÷ 7 =	43)	42 ÷ 7 =
12)	63 ÷ 7 =	28)	56 ÷ 7 =	44)	7 ÷ 7 =
13)	21 ÷ 7 =	29)	35 ÷ 7 =	45)	14 ÷ 7 =
14)	35 ÷ 7 =	30)	21 ÷ 7 =	46)	70 ÷ 7 =
15)	49 ÷ 7 =	31)	49 ÷ 7 =	47)	0 ÷ 7 =
16)	7 ÷ 7 =	32)	63 ÷ 7 =	48)	42 ÷ 7 =

DAY: 38

NAME: ____________ TIME: ___:___ SCORE: ______

DIVIDE 7

1)	7 ÷ 7 =	17)	49 ÷ 7 =	33)	63 ÷ 7 =
2)	14 ÷ 7 =	18)	0 ÷ 7 =	34)	14 ÷ 7 =
3)	70 ÷ 7 =	19)	56 ÷ 7 =	35)	56 ÷ 7 =
4)	7 ÷ 7 =	20)	28 ÷ 7 =	36)	28 ÷ 7 =
5)	63 ÷ 7 =	21)	56 ÷ 7 =	37)	21 ÷ 7 =
6)	21 ÷ 7 =	22)	35 ÷ 7 =	38)	70 ÷ 7 =
7)	42 ÷ 7 =	23)	21 ÷ 7 =	39)	7 ÷ 7 =
8)	49 ÷ 7 =	24)	49 ÷ 7 =	40)	35 ÷ 7 =
9)	7 ÷ 7 =	25)	14 ÷ 7 =	41)	42 ÷ 7 =
10)	56 ÷ 7 =	26)	21 ÷ 7 =	42)	35 ÷ 7 =
11)	70 ÷ 7 =	27)	7 ÷ 7 =	43)	28 ÷ 7 =
12)	63 ÷ 7 =	28)	14 ÷ 7 =	44)	42 ÷ 7 =
13)	35 ÷ 7 =	29)	28 ÷ 7 =	45)	14 ÷ 7 =
14)	70 ÷ 7 =	30)	35 ÷ 7 =	46)	42 ÷ 7 =
15)	49 ÷ 7 =	31)	49 ÷ 7 =	47)	0 ÷ 7 =
16)	42 ÷ 7 =	32)	14 ÷ 7 =	48)	28 ÷ 7 =

DAY: 39

NAME: ______ TIME: ___:___ SCORE: ______

DIVIDE 0 - 7

1)	42 ÷ 7 =	17)	4 ÷ 2 =	33)	15 ÷ 5 =
2)	18 ÷ 6 =	18)	16 ÷ 4 =	34)	12 ÷ 6 =
3)	54 ÷ 6 =	19)	36 ÷ 6 =	35)	49 ÷ 7 =
4)	27 ÷ 3 =	20)	21 ÷ 7 =	36)	40 ÷ 4 =
5)	36 ÷ 4 =	21)	0 ÷ 4 =	37)	14 ÷ 2 =
6)	0 ÷ 5 =	22)	63 ÷ 7 =	38)	70 ÷ 7 =
7)	9 ÷ 3 =	23)	50 ÷ 5 =	39)	21 ÷ 3 =
8)	10 ÷ 2 =	24)	35 ÷ 7 =	40)	6 ÷ 1 =
9)	60 ÷ 6 =	25)	20 ÷ 2 =	41)	14 ÷ 7 =
10)	32 ÷ 4 =	26)	12 ÷ 4 =	42)	42 ÷ 6 =
11)	45 ÷ 5 =	27)	4 ÷ 4 =	43)	20 ÷ 5 =
12)	0 ÷ 2 =	28)	25 ÷ 5 =	44)	16 ÷ 2 =
13)	2 ÷ 2 =	29)	18 ÷ 3 =	45)	30 ÷ 3 =
14)	10 ÷ 1 =	30)	8 ÷ 4 =	46)	7 ÷ 7 =
15)	6 ÷ 3 =	31)	6 ÷ 2 =	47)	48 ÷ 6 =
16)	28 ÷ 7 =	32)	20 ÷ 4 =	48)	5 ÷ 5 =

DAY: 40

NAME: ______________________ TIME: ___ : ___ SCORE: ________

DIVIDE 0 - 7

1)	15 ÷ 5 =	17)	42 ÷ 7 =	33)	4 ÷ 2 =
2)	12 ÷ 6 =	18)	18 ÷ 6 =	34)	16 ÷ 4 =
3)	35 ÷ 7 =	19)	5 ÷ 1 =	35)	36 ÷ 6 =
4)	36 ÷ 4 =	20)	27 ÷ 3 =	36)	21 ÷ 7 =
5)	14 ÷ 2 =	21)	12 ÷ 2 =	37)	0 ÷ 4 =
6)	70 ÷ 7 =	22)	0 ÷ 5 =	38)	63 ÷ 7 =
7)	30 ÷ 5 =	23)	9 ÷ 3 =	39)	50 ÷ 5 =
8)	6 ÷ 2 =	24)	10 ÷ 2 =	40)	24 ÷ 3 =
9)	14 ÷ 7 =	25)	8 ÷ 2 =	41)	20 ÷ 2 =
10)	42 ÷ 6 =	26)	32 ÷ 4 =	42)	18 ÷ 3 =
11)	45 ÷ 5 =	27)	28 ÷ 4 =	43)	0 ÷ 1 =
12)	16 ÷ 2 =	28)	0 ÷ 2 =	44)	25 ÷ 5 =
13)	30 ÷ 3 =	29)	2 ÷ 2 =	45)	24 ÷ 6 =
14)	28 ÷ 7 =	30)	10 ÷ 1 =	46)	8 ÷ 4 =
15)	48 ÷ 6 =	31)	6 ÷ 3 =	47)	18 ÷ 2 =
16)	4 ÷ 4 =	32)	21 ÷ 3 =	48)	20 ÷ 4 =

DAY: 41

NAME: ______ TIME: ___ : ___ SCORE: ______

DIVIDE 8

1)	16 ÷ 8 =	17)	48 ÷ 8 =	33)	0 ÷ 8 =
2)	40 ÷ 8 =	18)	64 ÷ 8 =	34)	32 ÷ 8 =
3)	32 ÷ 8 =	19)	40 ÷ 8 =	35)	16 ÷ 8 =
4)	56 ÷ 8 =	20)	8 ÷ 8 =	36)	24 ÷ 8 =
5)	8 ÷ 8 =	21)	56 ÷ 8 =	37)	40 ÷ 8 =
6)	80 ÷ 8 =	22)	16 ÷ 8 =	38)	72 ÷ 8 =
7)	72 ÷ 8 =	23)	24 ÷ 8 =	39)	64 ÷ 8 =
8)	64 ÷ 8 =	24)	64 ÷ 8 =	40)	80 ÷ 8 =
9)	40 ÷ 8 =	25)	80 ÷ 8 =	41)	48 ÷ 8 =
10)	56 ÷ 8 =	26)	0 ÷ 8 =	42)	24 ÷ 8 =
11)	80 ÷ 8 =	27)	32 ÷ 8 =	43)	40 ÷ 8 =
12)	64 ÷ 8 =	28)	24 ÷ 8 =	44)	56 ÷ 8 =
13)	24 ÷ 8 =	29)	8 ÷ 8 =	45)	72 ÷ 8 =
14)	32 ÷ 8 =	30)	72 ÷ 8 =	46)	80 ÷ 8 =
15)	48 ÷ 8 =	31)	48 ÷ 8 =	47)	16 ÷ 8 =
16)	8 ÷ 8 =	32)	56 ÷ 8 =	48)	72 ÷ 8 =

DAY: 42

NAME: ______________________ TIME: ____ : ____ SCORE: ________

DIVIDE 8

1)	24 ÷ 8 =	17)	8 ÷ 8 =	33)	48 ÷ 8 =
2)	40 ÷ 8 =	18)	32 ÷ 8 =	34)	64 ÷ 8 =
3)	72 ÷ 8 =	19)	16 ÷ 8 =	35)	40 ÷ 8 =
4)	64 ÷ 8 =	20)	8 ÷ 8 =	36)	16 ÷ 8 =
5)	80 ÷ 8 =	21)	56 ÷ 8 =	37)	40 ÷ 8 =
6)	0 ÷ 8 =	22)	16 ÷ 8 =	38)	32 ÷ 8 =
7)	24 ÷ 8 =	23)	24 ÷ 8 =	39)	56 ÷ 8 =
8)	64 ÷ 8 =	24)	64 ÷ 8 =	40)	8 ÷ 8 =
9)	40 ÷ 8 =	25)	80 ÷ 8 =	41)	80 ÷ 8 =
10)	56 ÷ 8 =	26)	48 ÷ 8 =	42)	72 ÷ 8 =
11)	80 ÷ 8 =	27)	32 ÷ 8 =	43)	40 ÷ 8 =
12)	64 ÷ 8 =	28)	24 ÷ 8 =	44)	56 ÷ 8 =
13)	24 ÷ 8 =	29)	8 ÷ 8 =	45)	72 ÷ 8 =
14)	32 ÷ 8 =	30)	72 ÷ 8 =	46)	80 ÷ 8 =
15)	0 ÷ 8 =	31)	48 ÷ 8 =	47)	16 ÷ 8 =
16)	8 ÷ 8 =	32)	56 ÷ 8 =	48)	72 ÷ 8 =

DAY: 43

NAME: ______________________ TIME: ____ : ____ SCORE: ________

DIVIDE 8

1)	24 ÷ 8 =	17)	8 ÷ 8 =	33)	48 ÷ 8 =
2)	64 ÷ 8 =	18)	32 ÷ 8 =	34)	40 ÷ 8 =
3)	40 ÷ 8 =	19)	24 ÷ 8 =	35)	72 ÷ 8 =
4)	16 ÷ 8 =	20)	64 ÷ 8 =	36)	64 ÷ 8 =
5)	40 ÷ 8 =	21)	40 ÷ 8 =	37)	80 ÷ 8 =
6)	32 ÷ 8 =	22)	56 ÷ 8 =	38)	16 ÷ 8 =
7)	16 ÷ 8 =	23)	80 ÷ 8 =	39)	56 ÷ 8 =
8)	8 ÷ 8 =	24)	64 ÷ 8 =	40)	8 ÷ 8 =
9)	56 ÷ 8 =	25)	0 ÷ 8 =	41)	80 ÷ 8 =
10)	16 ÷ 8 =	26)	16 ÷ 8 =	42)	72 ÷ 8 =
11)	24 ÷ 8 =	27)	32 ÷ 8 =	43)	40 ÷ 8 =
12)	64 ÷ 8 =	28)	24 ÷ 8 =	44)	56 ÷ 8 =
13)	24 ÷ 8 =	29)	8 ÷ 8 =	45)	72 ÷ 8 =
14)	32 ÷ 8 =	30)	72 ÷ 8 =	46)	80 ÷ 8 =
15)	16 ÷ 8 =	31)	48 ÷ 8 =	47)	0 ÷ 8 =
16)	48 ÷ 8 =	32)	56 ÷ 8 =	48)	72 ÷ 8 =

DAY: 44

NAME: ______________________ TIME: ___ : ___ SCORE: ________

DIVIDE 8

1)	16 ÷ 8 =	17)	48 ÷ 8 =	33)	64 ÷ 8 =
2)	40 ÷ 8 =	18)	64 ÷ 8 =	34)	40 ÷ 8 =
3)	32 ÷ 8 =	19)	80 ÷ 8 =	35)	56 ÷ 8 =
4)	56 ÷ 8 =	20)	0 ÷ 8 =	36)	80 ÷ 8 =
5)	8 ÷ 8 =	21)	24 ÷ 8 =	37)	40 ÷ 8 =
6)	80 ÷ 8 =	22)	40 ÷ 8 =	38)	72 ÷ 8 =
7)	72 ÷ 8 =	23)	24 ÷ 8 =	39)	64 ÷ 8 =
8)	8 ÷ 8 =	24)	64 ÷ 8 =	40)	40 ÷ 8 =
9)	32 ÷ 8 =	25)	80 ÷ 8 =	41)	8 ÷ 8 =
10)	16 ÷ 8 =	26)	16 ÷ 8 =	42)	56 ÷ 8 =
11)	24 ÷ 8 =	27)	32 ÷ 8 =	43)	16 ÷ 8 =
12)	64 ÷ 8 =	28)	24 ÷ 8 =	44)	48 ÷ 8 =
13)	24 ÷ 8 =	29)	0 ÷ 8 =	45)	72 ÷ 8 =
14)	32 ÷ 8 =	30)	72 ÷ 8 =	46)	80 ÷ 8 =
15)	16 ÷ 8 =	31)	48 ÷ 8 =	47)	16 ÷ 8 =
16)	8 ÷ 8 =	32)	56 ÷ 8 =	48)	72 ÷ 8 =

DAY: 45

NAME: ____________________ TIME: ___ : ___ SCORE: ________

DIVIDE 0 - 8

1)	63 ÷ 7 =	17)	12 ÷ 4 =	33)	16 ÷ 8 =
2)	48 ÷ 6 =	18)	15 ÷ 5 =	34)	72 ÷ 8 =
3)	56 ÷ 8 =	19)	54 ÷ 6 =	35)	35 ÷ 7 =
4)	15 ÷ 3 =	20)	21 ÷ 3 =	36)	36 ÷ 4 =
5)	12 ÷ 2 =	21)	0 ÷ 8 =	37)	45 ÷ 5 =
6)	0 ÷ 5 =	22)	56 ÷ 7 =	38)	70 ÷ 7 =
7)	9 ÷ 3 =	23)	40 ÷ 8 =	39)	27 ÷ 3 =
8)	20 ÷ 2 =	24)	7 ÷ 7 =	40)	36 ÷ 6 =
9)	18 ÷ 6 =	25)	40 ÷ 4 =	41)	9 ÷ 1 =
10)	32 ÷ 4 =	26)	8 ÷ 2 =	42)	24 ÷ 8 =
11)	6 ÷ 1 =	27)	20 ÷ 5 =	43)	21 ÷ 7 =
12)	1 ÷ 1 =	28)	25 ÷ 5 =	44)	16 ÷ 4 =
13)	18 ÷ 2 =	29)	30 ÷ 6 =	45)	80 ÷ 8 =
14)	64 ÷ 8 =	30)	10 ÷ 5 =	46)	42 ÷ 6 =
15)	42 ÷ 7 =	31)	16 ÷ 2 =	47)	48 ÷ 8 =
16)	28 ÷ 4 =	32)	24 ÷ 3 =	48)	8 ÷ 4 =

DAY: 46

NAME: ______________________ TIME: ____ : ____ SCORE: ________

DIVIDE 0 - 8

1)	12 ÷ 4 =	17)	15 ÷ 3 =	33)	49 ÷ 7 =
2)	15 ÷ 5 =	18)	72 ÷ 8 =	34)	48 ÷ 6 =
3)	54 ÷ 6 =	19)	35 ÷ 7 =	35)	56 ÷ 8 =
4)	21 ÷ 3 =	20)	36 ÷ 4 =	36)	35 ÷ 5 =
5)	0 ÷ 8 =	21)	50 ÷ 5 =	37)	12 ÷ 6 =
6)	56 ÷ 7 =	22)	63 ÷ 7 =	38)	0 ÷ 5 =
7)	40 ÷ 8 =	23)	27 ÷ 3 =	39)	24 ÷ 6 =
8)	7 ÷ 7 =	24)	36 ÷ 6 =	40)	20 ÷ 2 =
9)	40 ÷ 4 =	25)	3 ÷ 1 =	41)	18 ÷ 6 =
10)	12 ÷ 2 =	26)	24 ÷ 8 =	42)	32 ÷ 4 =
11)	20 ÷ 5 =	27)	21 ÷ 7 =	43)	8 ÷ 2 =
12)	25 ÷ 5 =	28)	16 ÷ 4 =	44)	1 ÷ 1 =
13)	14 ÷ 2 =	29)	80 ÷ 8 =	45)	18 ÷ 2 =
14)	9 ÷ 3 =	30)	28 ÷ 4 =	46)	64 ÷ 8 =
15)	16 ÷ 2 =	31)	48 ÷ 8 =	47)	42 ÷ 7 =
16)	24 ÷ 4 =	32)	8 ÷ 4 =	48)	40 ÷ 5 =

DAY: 47

NAME: ________________ TIME: ____ : ____ SCORE: ________

DIVIDE 9

1)	18 ÷ 9 =	17)	54 ÷ 9 =	33)	9 ÷ 9 =
2)	45 ÷ 9 =	18)	63 ÷ 9 =	34)	36 ÷ 9 =
3)	36 ÷ 9 =	19)	27 ÷ 9 =	35)	18 ÷ 9 =
4)	9 ÷ 9 =	20)	9 ÷ 9 =	36)	27 ÷ 9 =
5)	81 ÷ 9 =	21)	54 ÷ 9 =	37)	81 ÷ 9 =
6)	90 ÷ 9 =	22)	18 ÷ 9 =	38)	72 ÷ 9 =
7)	72 ÷ 9 =	23)	9 ÷ 9 =	39)	63 ÷ 9 =
8)	63 ÷ 9 =	24)	63 ÷ 9 =	40)	90 ÷ 9 =
9)	45 ÷ 9 =	25)	36 ÷ 9 =	41)	27 ÷ 9 =
10)	54 ÷ 9 =	26)	81 ÷ 9 =	42)	18 ÷ 9 =
11)	90 ÷ 9 =	27)	36 ÷ 9 =	43)	45 ÷ 9 =
12)	72 ÷ 9 =	28)	27 ÷ 9 =	44)	54 ÷ 9 =
13)	81 ÷ 9 =	29)	90 ÷ 9 =	45)	0 ÷ 9 =
14)	36 ÷ 9 =	30)	72 ÷ 9 =	46)	90 ÷ 9 =
15)	18 ÷ 9 =	31)	45 ÷ 9 =	47)	81 ÷ 9 =
16)	27 ÷ 9 =	32)	54 ÷ 9 =	48)	72 ÷ 9 =

DAY: 48

NAME: ____________________ TIME: ____ : ____ SCORE: ________

DIVIDE 9

1)	18 ÷ 9 =	17)	54 ÷ 9 =	33)	9 ÷ 9 =
2)	27 ÷ 9 =	18)	63 ÷ 9 =	34)	36 ÷ 9 =
3)	81 ÷ 9 =	19)	90 ÷ 9 =	35)	18 ÷ 9 =
4)	72 ÷ 9 =	20)	27 ÷ 9 =	36)	45 ÷ 9 =
5)	63 ÷ 9 =	21)	18 ÷ 9 =	37)	36 ÷ 9 =
6)	90 ÷ 9 =	22)	45 ÷ 9 =	38)	9 ÷ 9 =
7)	72 ÷ 9 =	23)	54 ÷ 9 =	39)	0 ÷ 9 =
8)	63 ÷ 9 =	24)	63 ÷ 9 =	40)	27 ÷ 9 =
9)	45 ÷ 9 =	25)	36 ÷ 9 =	41)	9 ÷ 9 =
10)	54 ÷ 9 =	26)	81 ÷ 9 =	42)	54 ÷ 9 =
11)	90 ÷ 9 =	27)	36 ÷ 9 =	43)	18 ÷ 9 =
12)	72 ÷ 9 =	28)	27 ÷ 9 =	44)	9 ÷ 9 =
13)	81 ÷ 9 =	29)	90 ÷ 9 =	45)	90 ÷ 9 =
14)	36 ÷ 9 =	30)	72 ÷ 9 =	46)	63 ÷ 9 =
15)	18 ÷ 9 =	31)	45 ÷ 9 =	47)	81 ÷ 9 =
16)	27 ÷ 9 =	32)	54 ÷ 9 =	48)	72 ÷ 9 =

DAY: 49

NAME: ____________ TIME: ___:___ SCORE: ______

DIVIDE 9

1)	36 ÷ 9 =	17)	54 ÷ 9 =	33)	36 ÷ 9 =
2)	27 ÷ 9 =	18)	63 ÷ 9 =	34)	81 ÷ 9 =
3)	90 ÷ 9 =	19)	90 ÷ 9 =	35)	18 ÷ 9 =
4)	72 ÷ 9 =	20)	27 ÷ 9 =	36)	27 ÷ 9 =
5)	18 ÷ 9 =	21)	18 ÷ 9 =	37)	81 ÷ 9 =
6)	90 ÷ 9 =	22)	45 ÷ 9 =	38)	72 ÷ 9 =
7)	72 ÷ 9 =	23)	54 ÷ 9 =	39)	63 ÷ 9 =
8)	63 ÷ 9 =	24)	63 ÷ 9 =	40)	54 ÷ 9 =
9)	45 ÷ 9 =	25)	9 ÷ 9 =	41)	9 ÷ 9 =
10)	54 ÷ 9 =	26)	36 ÷ 9 =	42)	45 ÷ 9 =
11)	90 ÷ 9 =	27)	18 ÷ 9 =	43)	18 ÷ 9 =
12)	72 ÷ 9 =	28)	45 ÷ 9 =	44)	63 ÷ 9 =
13)	81 ÷ 9 =	29)	36 ÷ 9 =	45)	9 ÷ 9 =
14)	36 ÷ 9 =	30)	9 ÷ 9 =	46)	90 ÷ 9 =
15)	0 ÷ 9 =	31)	81 ÷ 9 =	47)	81 ÷ 9 =
16)	27 ÷ 9 =	32)	27 ÷ 9 =	48)	72 ÷ 9 =

DAY: 50

NAME: ______________________ TIME: ____ : ____ SCORE: ________

DIVIDE 9

1)	54 ÷ 9 =	17)	27 ÷ 9 =	33)	45 ÷ 9 =
2)	0 ÷ 9 =	18)	81 ÷ 9 =	34)	63 ÷ 9 =
3)	72 ÷ 9 =	19)	18 ÷ 9 =	35)	90 ÷ 9 =
4)	63 ÷ 9 =	20)	36 ÷ 9 =	36)	27 ÷ 9 =
5)	18 ÷ 9 =	21)	9 ÷ 9 =	37)	18 ÷ 9 =
6)	90 ÷ 9 =	22)	27 ÷ 9 =	38)	45 ÷ 9 =
7)	72 ÷ 9 =	23)	36 ÷ 9 =	39)	54 ÷ 9 =
8)	63 ÷ 9 =	24)	54 ÷ 9 =	40)	63 ÷ 9 =
9)	45 ÷ 9 =	25)	90 ÷ 9 =	41)	9 ÷ 9 =
10)	54 ÷ 9 =	26)	54 ÷ 9 =	42)	36 ÷ 9 =
11)	90 ÷ 9 =	27)	18 ÷ 9 =	43)	18 ÷ 9 =
12)	72 ÷ 9 =	28)	81 ÷ 9 =	44)	45 ÷ 9 =
13)	81 ÷ 9 =	29)	27 ÷ 9 =	45)	36 ÷ 9 =
14)	36 ÷ 9 =	30)	72 ÷ 9 =	46)	9 ÷ 9 =
15)	18 ÷ 9 =	31)	81 ÷ 9 =	47)	81 ÷ 9 =
16)	27 ÷ 9 =	32)	63 ÷ 9 =	48)	27 ÷ 9 =

DAY: 51

NAME: ______________________ TIME: ____ : ____ SCORE: ________

DIVIDE 0 - 9

1)	72 ÷ 8 =	17)	12 ÷ 2 =	33)	14 ÷ 7 =
2)	42 ÷ 7 =	18)	25 ÷ 5 =	34)	72 ÷ 9 =
3)	63 ÷ 9 =	19)	54 ÷ 9 =	35)	35 ÷ 5 =
4)	15 ÷ 5 =	20)	21 ÷ 7 =	36)	4 ÷ 1 =
5)	12 ÷ 4 =	21)	0 ÷ 2 =	37)	50 ÷ 5 =
6)	0 ÷ 9 =	22)	48 ÷ 6 =	38)	18 ÷ 9 =
7)	9 ÷ 9 =	23)	40 ÷ 5 =	39)	6 ÷ 3 =
8)	20 ÷ 5 =	24)	12 ÷ 3 =	40)	14 ÷ 2 =
9)	18 ÷ 3 =	25)	40 ÷ 8 =	41)	5 ÷ 1 =
10)	32 ÷ 8 =	26)	36 ÷ 4 =	42)	64 ÷ 8 =
11)	0 ÷ 6 =	27)	0 ÷ 8 =	43)	56 ÷ 7 =
12)	54 ÷ 6 =	28)	36 ÷ 6 =	44)	20 ÷ 4 =
13)	16 ÷ 4 =	29)	8 ÷ 2 =	45)	60 ÷ 6 =
14)	81 ÷ 9 =	30)	21 ÷ 3 =	46)	28 ÷ 7 =
15)	42 ÷ 6 =	31)	16 ÷ 8 =	47)	18 ÷ 2 =
16)	28 ÷ 4 =	32)	24 ÷ 3 =	48)	28 ÷ 4 =

DAY: 52

NAME: ______________________ TIME: ____ : ____ SCORE: ________

DIVIDE 0 - 9

1)	15 ÷ 5 =	17)	12 ÷ 2 =	33)	0 ÷ 8 =
2)	72 ÷ 9 =	18)	25 ÷ 5 =	34)	42 ÷ 7 =
3)	35 ÷ 5 =	19)	54 ÷ 9 =	35)	63 ÷ 9 =
4)	32 ÷ 8 =	20)	21 ÷ 7 =	36)	5 ÷ 1 =
5)	50 ÷ 5 =	21)	0 ÷ 2 =	37)	12 ÷ 4 =
6)	18 ÷ 9 =	22)	48 ÷ 6 =	38)	0 ÷ 9 =
7)	6 ÷ 3 =	23)	40 ÷ 5 =	39)	9 ÷ 9 =
8)	54 ÷ 6 =	24)	12 ÷ 3 =	40)	24 ÷ 6 =
9)	7 ÷ 1 =	25)	40 ÷ 8 =	41)	15 ÷ 3 =
10)	64 ÷ 8 =	26)	36 ÷ 4 =	42)	24 ÷ 8 =
11)	56 ÷ 7 =	27)	72 ÷ 8 =	43)	0 ÷ 6 =
12)	16 ÷ 4 =	28)	20 ÷ 5 =	44)	49 ÷ 7 =
13)	60 ÷ 6 =	29)	18 ÷ 3 =	45)	20 ÷ 4 =
14)	28 ÷ 7 =	30)	21 ÷ 3 =	46)	81 ÷ 9 =
15)	8 ÷ 2 =	31)	16 ÷ 8 =	47)	42 ÷ 6 =
16)	28 ÷ 4 =	32)	24 ÷ 3 =	48)	32 ÷ 4 =

DAY: 53

NAME: ____________ TIME: ____ : ____ SCORE: ________

DIVIDE 10

1)	80 ÷ 10 =	17)	50 ÷ 10 =	33)	90 ÷ 10 =
2)	40 ÷ 10 =	18)	60 ÷ 10 =	34)	80 ÷ 10 =
3)	30 ÷ 10 =	19)	20 ÷ 10 =	35)	50 ÷ 10 =
4)	10 ÷ 10 =	20)	0 ÷ 10 =	36)	70 ÷ 10 =
5)	80 ÷ 10 =	21)	50 ÷ 10 =	37)	90 ÷ 10 =
6)	100 ÷ 10 =	22)	10 ÷ 10 =	38)	30 ÷ 10 =
7)	70 ÷ 10 =	23)	100 ÷ 10 =	39)	40 ÷ 10 =
8)	60 ÷ 10 =	24)	60 ÷ 10 =	40)	100 ÷ 10 =
9)	40 ÷ 10 =	25)	30 ÷ 10 =	41)	50 ÷ 10 =
10)	10 ÷ 10 =	26)	0 ÷ 10 =	42)	20 ÷ 10 =
11)	100 ÷ 10 =	27)	30 ÷ 10 =	43)	40 ÷ 10 =
12)	70 ÷ 10 =	28)	20 ÷ 10 =	44)	50 ÷ 10 =
13)	90 ÷ 10 =	29)	100 ÷ 10 =	45)	10 ÷ 10 =
14)	30 ÷ 10 =	30)	70 ÷ 10 =	46)	100 ÷ 10 =
15)	10 ÷ 10 =	31)	60 ÷ 10 =	47)	80 ÷ 10 =
16)	20 ÷ 10 =	32)	0 ÷ 10 =	48)	70 ÷ 10 =

DAY: 54

NAME: ____________________ TIME: ____ : ____ SCORE: ________

DIVIDE 10

1)	50 ÷ 10 =	17)	60 ÷ 10 =	33)	0 ÷ 10 =
2)	10 ÷ 10 =	18)	40 ÷ 10 =	34)	100 ÷ 10 =
3)	40 ÷ 10 =	19)	90 ÷ 10 =	35)	60 ÷ 10 =
4)	60 ÷ 10 =	20)	80 ÷ 10 =	36)	30 ÷ 10 =
5)	90 ÷ 10 =	21)	10 ÷ 10 =	37)	80 ÷ 10 =
6)	40 ÷ 10 =	22)	70 ÷ 10 =	38)	50 ÷ 10 =
7)	30 ÷ 10 =	23)	60 ÷ 10 =	39)	70 ÷ 10 =
8)	10 ÷ 10 =	24)	20 ÷ 10 =	40)	20 ÷ 10 =
9)	40 ÷ 10 =	25)	10 ÷ 10 =	41)	70 ÷ 10 =
10)	50 ÷ 10 =	26)	50 ÷ 10 =	42)	100 ÷ 10 =
11)	100 ÷ 10 =	27)	30 ÷ 10 =	43)	60 ÷ 10 =
12)	70 ÷ 10 =	28)	20 ÷ 10 =	44)	50 ÷ 10 =
13)	80 ÷ 10 =	29)	100 ÷ 10 =	45)	40 ÷ 10 =
14)	30 ÷ 10 =	30)	70 ÷ 10 =	46)	90 ÷ 10 =
15)	10 ÷ 10 =	31)	40 ÷ 10 =	47)	80 ÷ 10 =
16)	20 ÷ 10 =	32)	50 ÷ 10 =	48)	70 ÷ 10 =

DAY: 55

NAME: ________________ TIME: ____:____ SCORE: ________

DIVIDE 10

1)	20 ÷ 10 =	17)	0 ÷ 10 =	33)	90 ÷ 10 =
2)	100 ÷ 10 =	18)	80 ÷ 10 =	34)	60 ÷ 10 =
3)	70 ÷ 10 =	19)	50 ÷ 10 =	35)	20 ÷ 10 =
4)	60 ÷ 10 =	20)	70 ÷ 10 =	36)	0 ÷ 10 =
5)	80 ÷ 10 =	21)	80 ÷ 10 =	37)	50 ÷ 10 =
6)	40 ÷ 10 =	22)	10 ÷ 10 =	38)	30 ÷ 10 =
7)	30 ÷ 10 =	23)	100 ÷ 10 =	39)	40 ÷ 10 =
8)	10 ÷ 10 =	24)	60 ÷ 10 =	40)	100 ÷ 10 =
9)	40 ÷ 10 =	25)	30 ÷ 10 =	41)	50 ÷ 10 =
10)	90 ÷ 10 =	26)	80 ÷ 10 =	42)	20 ÷ 10 =
11)	100 ÷ 10 =	27)	30 ÷ 10 =	43)	40 ÷ 10 =
12)	70 ÷ 10 =	28)	20 ÷ 10 =	44)	50 ÷ 10 =
13)	80 ÷ 10 =	29)	100 ÷ 10 =	45)	10 ÷ 10 =
14)	30 ÷ 10 =	30)	70 ÷ 10 =	46)	90 ÷ 10 =
15)	10 ÷ 10 =	31)	40 ÷ 10 =	47)	80 ÷ 10 =
16)	20 ÷ 10 =	32)	50 ÷ 10 =	48)	60 ÷ 10 =

DAY: 56

NAME: ____________________ TIME: ____ : ____ SCORE: ________

DIVIDE 10

1)	50 ÷ 10 =	17)	0 ÷ 10 =	33)	10 ÷ 10 =
2)	10 ÷ 10 =	18)	80 ÷ 10 =	34)	100 ÷ 10 =
3)	40 ÷ 10 =	19)	20 ÷ 10 =	35)	60 ÷ 10 =
4)	60 ÷ 10 =	20)	70 ÷ 10 =	36)	30 ÷ 10 =
5)	80 ÷ 10 =	21)	20 ÷ 10 =	37)	80 ÷ 10 =
6)	40 ÷ 10 =	22)	90 ÷ 10 =	38)	60 ÷ 10 =
7)	30 ÷ 10 =	23)	60 ÷ 10 =	39)	40 ÷ 10 =
8)	10 ÷ 10 =	24)	20 ÷ 10 =	40)	100 ÷ 10 =
9)	40 ÷ 10 =	25)	10 ÷ 10 =	41)	80 ÷ 10 =
10)	50 ÷ 10 =	26)	50 ÷ 10 =	42)	100 ÷ 10 =
11)	90 ÷ 10 =	27)	30 ÷ 10 =	43)	70 ÷ 10 =
12)	70 ÷ 10 =	28)	20 ÷ 10 =	44)	50 ÷ 10 =
13)	80 ÷ 10 =	29)	100 ÷ 10 =	45)	0 ÷ 10 =
14)	30 ÷ 10 =	30)	70 ÷ 10 =	46)	90 ÷ 10 =
15)	10 ÷ 10 =	31)	40 ÷ 10 =	47)	60 ÷ 10 =
16)	20 ÷ 10 =	32)	50 ÷ 10 =	48)	70 ÷ 10 =

DAY: 57

NAME: ____________________ TIME: ____ : ____ SCORE: ________

DIVIDE 0 - 10

1)	80 ÷ 10 =	17)	50 ÷ 10 =	33)	15 ÷ 3 =
2)	42 ÷ 6 =	18)	50 ÷ 5 =	34)	100 ÷ 10 =
3)	63 ÷ 7 =	19)	54 ÷ 6 =	35)	35 ÷ 5 =
4)	10 ÷ 10 =	20)	21 ÷ 3 =	36)	28 ÷ 4 =
5)	20 ÷ 4 =	21)	2 ÷ 2 =	37)	20 ÷ 10 =
6)	81 ÷ 9 =	22)	32 ÷ 8 =	38)	16 ÷ 2 =
7)	56 ÷ 8 =	23)	40 ÷ 10 =	39)	6 ÷ 6 =
8)	20 ÷ 5 =	24)	12 ÷ 6 =	40)	0 ÷ 3 =
9)	18 ÷ 6 =	25)	40 ÷ 5 =	41)	9 ÷ 1 =
10)	32 ÷ 4 =	26)	36 ÷ 9 =	42)	64 ÷ 8 =
11)	0 ÷ 1 =	27)	54 ÷ 9 =	43)	56 ÷ 7 =
12)	24 ÷ 3 =	28)	25 ÷ 5 =	44)	24 ÷ 4 =
13)	16 ÷ 8 =	29)	12 ÷ 3 =	45)	30 ÷ 6 =
14)	90 ÷ 9 =	30)	21 ÷ 7 =	46)	27 ÷ 3 =
15)	42 ÷ 7 =	31)	48 ÷ 8 =	47)	12 ÷ 2 =
16)	18 ÷ 2 =	32)	27 ÷ 9 =	48)	45 ÷ 5 =

DAY: 58

NAME: ______________________ TIME: ____ : ____ SCORE: ________

DIVIDE 0 - 10

1)	15 ÷ 3 =	17)	50 ÷ 10 =	33)	80 ÷ 10 =
2)	100 ÷ 10 =	18)	30 ÷ 5 =	34)	42 ÷ 6 =
3)	35 ÷ 5 =	19)	54 ÷ 6 =	35)	28 ÷ 7 =
4)	32 ÷ 4 =	20)	21 ÷ 3 =	36)	10 ÷ 10 =
5)	20 ÷ 10 =	21)	2 ÷ 2 =	37)	20 ÷ 4 =
6)	18 ÷ 2 =	22)	48 ÷ 8 =	38)	81 ÷ 9 =
7)	6 ÷ 6 =	23)	40 ÷ 10 =	39)	56 ÷ 8 =
8)	54 ÷ 9 =	24)	12 ÷ 6 =	40)	20 ÷ 5 =
9)	1 ÷ 1 =	25)	40 ÷ 5 =	41)	18 ÷ 6 =
10)	64 ÷ 8 =	26)	36 ÷ 9 =	42)	36 ÷ 4 =
11)	63 ÷ 7 =	27)	90 ÷ 9 =	43)	0 ÷ 2 =
12)	24 ÷ 4 =	28)	25 ÷ 5 =	44)	24 ÷ 3 =
13)	30 ÷ 6 =	29)	12 ÷ 3 =	45)	16 ÷ 8 =
14)	27 ÷ 3 =	30)	21 ÷ 7 =	46)	72 ÷ 9 =
15)	12 ÷ 2 =	31)	72 ÷ 8 =	47)	42 ÷ 7 =
16)	45 ÷ 5 =	32)	27 ÷ 9 =	48)	16 ÷ 2 =

DAY: 59

NAME: ______________________ TIME: ___ : ___ SCORE: ________

DIVIDE 11

1)	88 ÷ 11 =	17)	55 ÷ 11 =	33)	99 ÷ 11 =
2)	44 ÷ 11 =	18)	66 ÷ 11 =	34)	88 ÷ 11 =
3)	33 ÷ 11 =	19)	22 ÷ 11 =	35)	0 ÷ 11 =
4)	66 ÷ 11 =	20)	11 ÷ 11 =	36)	77 ÷ 11 =
5)	0 ÷ 11 =	21)	55 ÷ 11 =	37)	99 ÷ 11 =
6)	121 ÷ 11 =	22)	22 ÷ 11 =	38)	33 ÷ 11 =
7)	77 ÷ 11 =	23)	110 ÷ 11 =	39)	44 ÷ 11 =
8)	66 ÷ 11 =	24)	66 ÷ 11 =	40)	110 ÷ 11 =
9)	44 ÷ 11 =	25)	33 ÷ 11 =	41)	55 ÷ 11 =
10)	55 ÷ 11 =	26)	88 ÷ 11 =	42)	0 ÷ 11 =
11)	110 ÷ 11 =	27)	33 ÷ 11 =	43)	44 ÷ 11 =
12)	77 ÷ 11 =	28)	22 ÷ 11 =	44)	55 ÷ 11 =
13)	99 ÷ 11 =	29)	110 ÷ 11 =	45)	11 ÷ 11 =
14)	33 ÷ 11 =	30)	77 ÷ 11 =	46)	121 ÷ 11 =
15)	11 ÷ 11 =	31)	44 ÷ 11 =	47)	88 ÷ 11 =
16)	22 ÷ 11 =	32)	55 ÷ 11 =	48)	77 ÷ 11 =

DAY: 60

NAME: ______________________ TIME: ____ : ____ SCORE: ________

DIVIDE 11

1)	33 ÷ 11 =	17)	66 ÷ 11 =	33)	99 ÷ 11 =
2)	44 ÷ 11 =	18)	44 ÷ 11 =	34)	88 ÷ 11 =
3)	121 ÷ 11 =	19)	55 ÷ 11 =	35)	55 ÷ 11 =
4)	55 ÷ 11 =	20)	99 ÷ 11 =	36)	77 ÷ 11 =
5)	11 ÷ 11 =	21)	0 ÷ 11 =	37)	66 ÷ 11 =
6)	110 ÷ 11 =	22)	22 ÷ 11 =	38)	88 ÷ 11 =
7)	77 ÷ 11 =	23)	110 ÷ 11 =	39)	44 ÷ 11 =
8)	55 ÷ 11 =	24)	66 ÷ 11 =	40)	33 ÷ 11 =
9)	66 ÷ 11 =	25)	33 ÷ 11 =	41)	11 ÷ 11 =
10)	22 ÷ 11 =	26)	99 ÷ 11 =	42)	88 ÷ 11 =
11)	11 ÷ 11 =	27)	33 ÷ 11 =	43)	44 ÷ 11 =
12)	77 ÷ 11 =	28)	22 ÷ 11 =	44)	55 ÷ 11 =
13)	88 ÷ 11 =	29)	110 ÷ 11 =	45)	0 ÷ 11 =
14)	33 ÷ 11 =	30)	77 ÷ 11 =	46)	121 ÷ 11 =
15)	11 ÷ 11 =	31)	44 ÷ 11 =	47)	99 ÷ 11 =
16)	22 ÷ 11 =	32)	66 ÷ 11 =	48)	77 ÷ 11 =

DAY: 61

NAME: ______________________ TIME: ____ : ____ SCORE: ________

DIVIDE 11

1)	33 ÷ 11 =	17)	66 ÷ 11 =	33)	99 ÷ 11 =
2)	22 ÷ 11 =	18)	44 ÷ 11 =	34)	88 ÷ 11 =
3)	11 ÷ 11 =	19)	55 ÷ 11 =	35)	44 ÷ 11 =
4)	77 ÷ 11 =	20)	121 ÷ 11 =	36)	55 ÷ 11 =
5)	88 ÷ 11 =	21)	66 ÷ 11 =	37)	11 ÷ 11 =
6)	110 ÷ 11 =	22)	11 ÷ 11 =	38)	0 ÷ 11 =
7)	77 ÷ 11 =	23)	110 ÷ 11 =	39)	44 ÷ 11 =
8)	55 ÷ 11 =	24)	66 ÷ 11 =	40)	33 ÷ 11 =
9)	66 ÷ 11 =	25)	33 ÷ 11 =	41)	11 ÷ 11 =
10)	22 ÷ 11 =	26)	0 ÷ 11 =	42)	88 ÷ 11 =
11)	121 ÷ 11 =	27)	44 ÷ 11 =	43)	99 ÷ 11 =
12)	55 ÷ 11 =	28)	22 ÷ 11 =	44)	77 ÷ 11 =
13)	11 ÷ 11 =	29)	110 ÷ 11 =	45)	88 ÷ 11 =
14)	33 ÷ 11 =	30)	77 ÷ 11 =	46)	0 ÷ 11 =
15)	44 ÷ 11 =	31)	44 ÷ 11 =	47)	99 ÷ 11 =
16)	22 ÷ 11 =	32)	55 ÷ 11 =	48)	77 ÷ 11 =

DAY: 62

NAME: ______________________ TIME: ____:____ SCORE: ________

DIVIDE 11

1)	33 ÷ 11 =	17)	99 ÷ 11 =	33)	66 ÷ 11 =
2)	110 ÷ 11 =	18)	121 ÷ 11 =	34)	44 ÷ 11 =
3)	66 ÷ 11 =	19)	44 ÷ 11 =	35)	55 ÷ 11 =
4)	33 ÷ 11 =	20)	55 ÷ 11 =	36)	0 ÷ 11 =
5)	88 ÷ 11 =	21)	11 ÷ 11 =	37)	88 ÷ 11 =
6)	55 ÷ 11 =	22)	99 ÷ 11 =	38)	11 ÷ 11 =
7)	77 ÷ 11 =	23)	44 ÷ 11 =	39)	22 ÷ 11 =
8)	99 ÷ 11 =	24)	33 ÷ 11 =	40)	33 ÷ 11 =
9)	66 ÷ 11 =	25)	11 ÷ 11 =	41)	77 ÷ 11 =
10)	22 ÷ 11 =	26)	88 ÷ 11 =	42)	66 ÷ 11 =
11)	110 ÷ 11 =	27)	33 ÷ 11 =	43)	121 ÷ 11 =
12)	55 ÷ 11 =	28)	22 ÷ 11 =	44)	77 ÷ 11 =
13)	11 ÷ 11 =	29)	0 ÷ 11 =	45)	88 ÷ 11 =
14)	33 ÷ 11 =	30)	77 ÷ 11 =	46)	110 ÷ 11 =
15)	121 ÷ 11 =	31)	44 ÷ 11 =	47)	99 ÷ 11 =
16)	22 ÷ 11 =	32)	55 ÷ 11 =	48)	77 ÷ 11 =

DAY: 63

NAME: ____________ TIME: ___ : ___ SCORE: ______

DIVIDE 0 - 11

1)	90 ÷ 10 =	17)	50 ÷ 10 =	33)	15 ÷ 5 =
2)	48 ÷ 6 =	18)	44 ÷ 11 =	34)	99 ÷ 11 =
3)	66 ÷ 11 =	19)	63 ÷ 9 =	35)	35 ÷ 7 =
4)	72 ÷ 9 =	20)	21 ÷ 7 =	36)	0 ÷ 8 =
5)	20 ÷ 4 =	21)	2 ÷ 1 =	37)	20 ÷ 2 =
6)	48 ÷ 8 =	22)	36 ÷ 6 =	38)	18 ÷ 6 =
7)	56 ÷ 7 =	23)	40 ÷ 8 =	39)	6 ÷ 3 =
8)	18 ÷ 2 =	24)	12 ÷ 3 =	40)	56 ÷ 8 =
9)	18 ÷ 3 =	25)	40 ÷ 10 =	41)	54 ÷ 9 =
10)	32 ÷ 8 =	26)	36 ÷ 4 =	42)	49 ÷ 7 =
11)	2 ÷ 2 =	27)	54 ÷ 6 =	43)	0 ÷ 10 =
12)	24 ÷ 8 =	28)	25 ÷ 5 =	44)	121 ÷ 11 =
13)	16 ÷ 4 =	29)	12 ÷ 4 =	45)	30 ÷ 5 =
14)	40 ÷ 5 =	30)	21 ÷ 3 =	46)	27 ÷ 9 =
15)	42 ÷ 6 =	31)	81 ÷ 9 =	47)	10 ÷ 1 =
16)	18 ÷ 9 =	32)	24 ÷ 3 =	48)	45 ÷ 9 =

DAY: 64

NAME: ____________________ TIME: ____ : ____ SCORE: ________

DIVIDE 0 - 11

1)	$15 \div 5 =$	17)	$70 \div 10 =$	33)	$50 \div 10 =$
2)	$99 \div 11 =$	18)	$48 \div 6 =$	34)	$44 \div 11 =$
3)	$35 \div 7 =$	19)	$66 \div 11 =$	35)	$0 \div 9 =$
4)	$32 \div 8 =$	20)	$72 \div 9 =$	36)	$21 \div 7 =$
5)	$20 \div 2 =$	21)	$24 \div 4 =$	37)	$2 \div 1 =$
6)	$18 \div 6 =$	22)	$48 \div 8 =$	38)	$24 \div 6 =$
7)	$6 \div 3 =$	23)	$56 \div 7 =$	39)	$40 \div 8 =$
8)	$56 \div 8 =$	24)	$20 \div 4 =$	40)	$12 \div 3 =$
9)	$54 \div 9 =$	25)	$18 \div 3 =$	41)	$40 \div 10 =$
10)	$49 \div 7 =$	26)	$0 \div 8 =$	42)	$36 \div 4 =$
11)	$90 \div 10 =$	27)	$2 \div 2 =$	43)	$54 \div 6 =$
12)	$22 \div 11 =$	28)	$24 \div 8 =$	44)	$25 \div 5 =$
13)	$30 \div 5 =$	29)	$16 \div 4 =$	45)	$12 \div 4 =$
14)	$27 \div 9 =$	30)	$40 \div 5 =$	46)	$21 \div 3 =$
15)	$10 \div 1 =$	31)	$42 \div 6 =$	47)	$81 \div 9 =$
16)	$45 \div 9 =$	32)	$18 \div 9 =$	48)	$24 \div 3 =$

DAY: 65

NAME: ____________________ TIME: ___ : ___ SCORE: ________

DIVIDE 12

1)	96 ÷ 12 =	17)	144 ÷ 12 =	33)	108 ÷ 12 =
2)	48 ÷ 12 =	18)	72 ÷ 12 =	34)	96 ÷ 12 =
3)	36 ÷ 12 =	19)	24 ÷ 12 =	35)	0 ÷ 12 =
4)	12 ÷ 12 =	20)	12 ÷ 12 =	36)	84 ÷ 12 =
5)	96 ÷ 12 =	21)	60 ÷ 12 =	37)	96 ÷ 12 =
6)	120 ÷ 12 =	22)	108 ÷ 12 =	38)	36 ÷ 12 =
7)	84 ÷ 12 =	23)	132 ÷ 12 =	39)	48 ÷ 12 =
8)	72 ÷ 12 =	24)	72 ÷ 12 =	40)	120 ÷ 12 =
9)	144 ÷ 12 =	25)	0 ÷ 12 =	41)	60 ÷ 12 =
10)	60 ÷ 12 =	26)	96 ÷ 12 =	42)	12 ÷ 12 =
11)	120 ÷ 12 =	27)	36 ÷ 12 =	43)	48 ÷ 12 =
12)	84 ÷ 12 =	28)	24 ÷ 12 =	44)	108 ÷ 12 =
13)	96 ÷ 12 =	29)	120 ÷ 12 =	45)	12 ÷ 12 =
14)	36 ÷ 12 =	30)	84 ÷ 12 =	46)	132 ÷ 12 =
15)	108 ÷ 12 =	31)	48 ÷ 12 =	47)	36 ÷ 12 =
16)	24 ÷ 12 =	32)	60 ÷ 12 =	48)	72 ÷ 12 =

DAY: 66

NAME: ______________________ TIME: ___ : ___ SCORE: ______

DIVIDE 12

1)	144 ÷ 12 =	17)	72 ÷ 12 =	33)	108 ÷ 12 =
2)	60 ÷ 12 =	18)	24 ÷ 12 =	34)	96 ÷ 12 =
3)	12 ÷ 12 =	19)	60 ÷ 12 =	35)	60 ÷ 12 =
4)	48 ÷ 12 =	20)	120 ÷ 12 =	36)	84 ÷ 12 =
5)	132 ÷ 12 =	21)	84 ÷ 12 =	37)	0 ÷ 12 =
6)	120 ÷ 12 =	22)	12 ÷ 12 =	38)	36 ÷ 12 =
7)	84 ÷ 12 =	23)	120 ÷ 12 =	39)	48 ÷ 12 =
8)	60 ÷ 12 =	24)	72 ÷ 12 =	40)	96 ÷ 12 =
9)	72 ÷ 12 =	25)	36 ÷ 12 =	41)	48 ÷ 12 =
10)	24 ÷ 12 =	26)	132 ÷ 12 =	42)	36 ÷ 12 =
11)	12 ÷ 12 =	27)	108 ÷ 12 =	43)	12 ÷ 12 =
12)	60 ÷ 12 =	28)	24 ÷ 12 =	44)	0 ÷ 12 =
13)	96 ÷ 12 =	29)	120 ÷ 12 =	45)	144 ÷ 12 =
14)	36 ÷ 12 =	30)	84 ÷ 12 =	46)	120 ÷ 12 =
15)	108 ÷ 12 =	31)	48 ÷ 12 =	47)	96 ÷ 12 =
16)	24 ÷ 12 =	32)	60 ÷ 12 =	48)	72 ÷ 12 =

DAY: 67

NAME: ____________________ TIME: ___ : ___ SCORE: ________

DIVIDE 12

1)	120 ÷ 12 =	17)	72 ÷ 12 =	33)	60 ÷ 12 =
2)	108 ÷ 12 =	18)	132 ÷ 12 =	34)	24 ÷ 12 =
3)	96 ÷ 12 =	19)	48 ÷ 12 =	35)	48 ÷ 12 =
4)	60 ÷ 12 =	20)	36 ÷ 12 =	36)	0 ÷ 12 =
5)	84 ÷ 12 =	21)	12 ÷ 12 =	37)	108 ÷ 12 =
6)	144 ÷ 12 =	22)	60 ÷ 12 =	38)	36 ÷ 12 =
7)	96 ÷ 12 =	23)	120 ÷ 12 =	39)	48 ÷ 12 =
8)	60 ÷ 12 =	24)	72 ÷ 12 =	40)	120 ÷ 12 =
9)	72 ÷ 12 =	25)	36 ÷ 12 =	41)	132 ÷ 12 =
10)	24 ÷ 12 =	26)	96 ÷ 12 =	42)	108 ÷ 12 =
11)	12 ÷ 12 =	27)	0 ÷ 12 =	43)	84 ÷ 12 =
12)	60 ÷ 12 =	28)	24 ÷ 12 =	44)	12 ÷ 12 =
13)	96 ÷ 12 =	29)	120 ÷ 12 =	45)	72 ÷ 12 =
14)	36 ÷ 12 =	30)	84 ÷ 12 =	46)	120 ÷ 12 =
15)	12 ÷ 12 =	31)	48 ÷ 12 =	47)	144 ÷ 12 =
16)	24 ÷ 12 =	32)	72 ÷ 12 =	48)	84 ÷ 12 =

DAY: 68

NAME: ______________________ TIME: ____ : ____ SCORE: ________

DIVIDE 12

1)	120 ÷ 12 =	17)	72 ÷ 12 =	33)	12 ÷ 12 =
2)	60 ÷ 12 =	18)	96 ÷ 12 =	34)	24 ÷ 12 =
3)	96 ÷ 12 =	19)	48 ÷ 12 =	35)	48 ÷ 12 =
4)	108 ÷ 12 =	20)	36 ÷ 12 =	36)	0 ÷ 12 =
5)	144 ÷ 12 =	21)	12 ÷ 12 =	37)	132 ÷ 12 =
6)	120 ÷ 12 =	22)	96 ÷ 12 =	38)	120 ÷ 12 =
7)	84 ÷ 12 =	23)	0 ÷ 12 =	39)	48 ÷ 12 =
8)	60 ÷ 12 =	24)	72 ÷ 12 =	40)	72 ÷ 12 =
9)	72 ÷ 12 =	25)	36 ÷ 12 =	41)	60 ÷ 12 =
10)	24 ÷ 12 =	26)	96 ÷ 12 =	42)	36 ÷ 12 =
11)	12 ÷ 12 =	27)	132 ÷ 12 =	43)	84 ÷ 12 =
12)	60 ÷ 12 =	28)	24 ÷ 12 =	44)	108 ÷ 12 =
13)	96 ÷ 12 =	29)	120 ÷ 12 =	45)	0 ÷ 12 =
14)	36 ÷ 12 =	30)	84 ÷ 12 =	46)	144 ÷ 12 =
15)	12 ÷ 12 =	31)	48 ÷ 12 =	47)	72 ÷ 12 =
16)	24 ÷ 12 =	32)	60 ÷ 12 =	48)	84 ÷ 12 =

DAY: 69

NAME: ______________________ TIME: ____ : ____ SCORE: ________

DIVIDE 0 - 12

1)	24 ÷ 12 =	17)	55 ÷ 11 =	33)	15 ÷ 3 =
2)	48 ÷ 8 =	18)	48 ÷ 12 =	34)	11 ÷ 11 =
3)	88 ÷ 11 =	19)	54 ÷ 6 =	35)	35 ÷ 5 =
4)	72 ÷ 8 =	20)	21 ÷ 3 =	36)	32 ÷ 4 =
5)	16 ÷ 4 =	21)	4 ÷ 2 =	37)	20 ÷ 10 =
6)	48 ÷ 6 =	22)	28 ÷ 4 =	38)	18 ÷ 6 =
7)	56 ÷ 8 =	23)	40 ÷ 5 =	39)	6 ÷ 2 =
8)	36 ÷ 12 =	24)	6 ÷ 1 =	40)	56 ÷ 7 =
9)	80 ÷ 10 =	25)	36 ÷ 9 =	41)	72 ÷ 12 =
10)	32 ÷ 8 =	26)	30 ÷ 5 =	42)	49 ÷ 7 =
11)	27 ÷ 9 =	27)	0 ÷ 4 =	43)	90 ÷ 9 =
12)	12 ÷ 2 =	28)	90 ÷ 10 =	44)	132 ÷ 12 =
13)	18 ÷ 3 =	29)	12 ÷ 6 =	45)	36 ÷ 6 =
14)	4 ÷ 1 =	30)	21 ÷ 7 =	46)	16 ÷ 8 =
15)	42 ÷ 7 =	31)	55 ÷ 11 =	47)	10 ÷ 10 =
16)	18 ÷ 2 =	32)	60 ÷ 12 =	48)	40 ÷ 4 =

DAY: 70

NAME: ______________________ TIME: ____ : ____ SCORE: ________

DIVIDE 0 - 12

1)	15 ÷ 3 =	17)	55 ÷ 11 =	33)	24 ÷ 12 =
2)	11 ÷ 11 =	18)	48 ÷ 12 =	34)	0 ÷ 7 =
3)	35 ÷ 5 =	19)	54 ÷ 6 =	35)	33 ÷ 11 =
4)	32 ÷ 4 =	20)	21 ÷ 3 =	36)	72 ÷ 8 =
5)	81 ÷ 9 =	21)	4 ÷ 2 =	37)	20 ÷ 4 =
6)	18 ÷ 6 =	22)	48 ÷ 8 =	38)	48 ÷ 6 =
7)	6 ÷ 2 =	23)	40 ÷ 5 =	39)	36 ÷ 4 =
8)	56 ÷ 7 =	24)	1 ÷ 1 =	40)	144 ÷ 12 =
9)	72 ÷ 12 =	25)	36 ÷ 9 =	41)	80 ÷ 10 =
10)	49 ÷ 7 =	26)	30 ÷ 5 =	42)	32 ÷ 8 =
11)	100 ÷ 10 =	27)	56 ÷ 8 =	43)	20 ÷ 10 =
12)	72 ÷ 9 =	28)	90 ÷ 10 =	44)	12 ÷ 2 =
13)	36 ÷ 6 =	29)	12 ÷ 6 =	45)	27 ÷ 3 =
14)	16 ÷ 8 =	30)	21 ÷ 7 =	46)	4 ÷ 1 =
15)	10 ÷ 10 =	31)	121 ÷ 11 =	47)	42 ÷ 7 =
16)	45 ÷ 5 =	32)	60 ÷ 12 =	48)	18 ÷ 2 =

DAY: 71

NAME: ______________________ TIME: ____:____ SCORE: ________

DIVIDE 13

1)	104 ÷ 13 =	17)	65 ÷ 13 =	33)	117 ÷ 13 =
2)	52 ÷ 13 =	18)	78 ÷ 13 =	34)	104 ÷ 13 =
3)	39 ÷ 13 =	19)	26 ÷ 13 =	35)	65 ÷ 13 =
4)	0 ÷ 13 =	20)	117 ÷ 13 =	36)	143 ÷ 13 =
5)	104 ÷ 13 =	21)	65 ÷ 13 =	37)	39 ÷ 13 =
6)	130 ÷ 13 =	22)	13 ÷ 13 =	38)	26 ÷ 13 =
7)	91 ÷ 13 =	23)	130 ÷ 13 =	39)	156 ÷ 13 =
8)	78 ÷ 13 =	24)	169 ÷ 13 =	40)	130 ÷ 13 =
9)	52 ÷ 13 =	25)	39 ÷ 13 =	41)	65 ÷ 13 =
10)	65 ÷ 13 =	26)	0 ÷ 13 =	42)	169 ÷ 13 =
11)	130 ÷ 13 =	27)	39 ÷ 13 =	43)	52 ÷ 13 =
12)	91 ÷ 13 =	28)	26 ÷ 13 =	44)	0 ÷ 13 =
13)	104 ÷ 13 =	29)	143 ÷ 13 =	45)	13 ÷ 13 =
14)	156 ÷ 13 =	30)	91 ÷ 13 =	46)	117 ÷ 13 =
15)	13 ÷ 13 =	31)	52 ÷ 13 =	47)	104 ÷ 13 =
16)	26 ÷ 13 =	32)	78 ÷ 13 =	48)	91 ÷ 13 =

DAY: 72

NAME: ______________________ TIME: ____ : ____ SCORE: ________

DIVIDE 13

1)	91 ÷ 13 =	17)	0 ÷ 13 =	33)	143 ÷ 13 =
2)	39 ÷ 13 =	18)	78 ÷ 13 =	34)	104 ÷ 13 =
3)	52 ÷ 13 =	19)	26 ÷ 13 =	35)	65 ÷ 13 =
4)	130 ÷ 13 =	20)	117 ÷ 13 =	36)	91 ÷ 13 =
5)	39 ÷ 13 =	21)	65 ÷ 13 =	37)	104 ÷ 13 =
6)	104 ÷ 13 =	22)	13 ÷ 13 =	38)	52 ÷ 13 =
7)	117 ÷ 13 =	23)	130 ÷ 13 =	39)	39 ÷ 13 =
8)	26 ÷ 13 =	24)	143 ÷ 13 =	40)	13 ÷ 13 =
9)	0 ÷ 13 =	25)	104 ÷ 13 =	41)	169 ÷ 13 =
10)	65 ÷ 13 =	26)	156 ÷ 13 =	42)	13 ÷ 13 =
11)	130 ÷ 13 =	27)	91 ÷ 13 =	43)	52 ÷ 13 =
12)	169 ÷ 13 =	28)	78 ÷ 13 =	44)	65 ÷ 13 =
13)	78 ÷ 13 =	29)	130 ÷ 13 =	45)	13 ÷ 13 =
14)	39 ÷ 13 =	30)	26 ÷ 13 =	46)	156 ÷ 13 =
15)	13 ÷ 13 =	31)	52 ÷ 13 =	47)	0 ÷ 13 =
16)	26 ÷ 13 =	32)	78 ÷ 13 =	48)	91 ÷ 13 =

DAY: 73

NAME: ____________ TIME: ___:___ SCORE: ________

DIVIDE 13

1)	78 ÷ 13 =	17)	65 ÷ 13 =	33)	117 ÷ 13 =
2)	26 ÷ 13 =	18)	78 ÷ 13 =	34)	169 ÷ 13 =
3)	52 ÷ 13 =	19)	26 ÷ 13 =	35)	65 ÷ 13 =
4)	169 ÷ 13 =	20)	143 ÷ 13 =	36)	91 ÷ 13 =
5)	130 ÷ 13 =	21)	52 ÷ 13 =	37)	26 ÷ 13 =
6)	104 ÷ 13 =	22)	65 ÷ 13 =	38)	52 ÷ 13 =
7)	39 ÷ 13 =	23)	13 ÷ 13 =	39)	39 ÷ 13 =
8)	13 ÷ 13 =	24)	78 ÷ 13 =	40)	13 ÷ 13 =
9)	52 ÷ 13 =	25)	0 ÷ 13 =	41)	65 ÷ 13 =
10)	156 ÷ 13 =	26)	130 ÷ 13 =	42)	117 ÷ 13 =
11)	39 ÷ 13 =	27)	91 ÷ 13 =	43)	65 ÷ 13 =
12)	91 ÷ 13 =	28)	78 ÷ 13 =	44)	0 ÷ 13 =
13)	104 ÷ 13 =	29)	143 ÷ 13 =	45)	156 ÷ 13 =
14)	39 ÷ 13 =	30)	91 ÷ 13 =	46)	39 ÷ 13 =
15)	13 ÷ 13 =	31)	52 ÷ 13 =	47)	169 ÷ 13 =
16)	26 ÷ 13 =	32)	0 ÷ 13 =	48)	91 ÷ 13 =

DAY: 74

NAME: ______________________ TIME: ___:___ SCORE: ________

DIVIDE 13

1)	91 ÷ 13 =	17)	117 ÷ 13 =	33)	143 ÷ 13 =
2)	26 ÷ 13 =	18)	0 ÷ 13 =	34)	78 ÷ 13 =
3)	156 ÷ 13 =	19)	65 ÷ 13 =	35)	26 ÷ 13 =
4)	130 ÷ 13 =	20)	91 ÷ 13 =	36)	0 ÷ 13 =
5)	91 ÷ 13 =	21)	104 ÷ 13 =	37)	52 ÷ 13 =
6)	78 ÷ 13 =	22)	52 ÷ 13 =	38)	65 ÷ 13 =
7)	39 ÷ 13 =	23)	13 ÷ 13 =	39)	39 ÷ 13 =
8)	13 ÷ 13 =	24)	78 ÷ 13 =	40)	169 ÷ 13 =
9)	52 ÷ 13 =	25)	104 ÷ 13 =	41)	65 ÷ 13 =
10)	156 ÷ 13 =	26)	78 ÷ 13 =	42)	117 ÷ 13 =
11)	39 ÷ 13 =	27)	130 ÷ 13 =	43)	0 ÷ 13 =
12)	169 ÷ 13 =	28)	26 ÷ 13 =	44)	13 ÷ 13 =
13)	143 ÷ 13 =	29)	117 ÷ 13 =	45)	130 ÷ 13 =
14)	39 ÷ 13 =	30)	91 ÷ 13 =	46)	39 ÷ 13 =
15)	13 ÷ 13 =	31)	52 ÷ 13 =	47)	104 ÷ 13 =
16)	26 ÷ 13 =	32)	65 ÷ 13 =	48)	78 ÷ 13 =

DAY: 75

NAME: ______________________ TIME: ____ : ____ SCORE: ________

DIVIDE 0 - 13

1)	84 ÷ 12 =	17)	65 ÷ 13 =	33)	24 ÷ 3 =
2)	56 ÷ 7 =	18)	48 ÷ 6 =	34)	26 ÷ 13 =
3)	81 ÷ 9 =	19)	0 ÷ 9 =	35)	45 ÷ 9 =
4)	70 ÷ 10 =	20)	121 ÷ 11 =	36)	36 ÷ 12 =
5)	28 ÷ 4 =	21)	14 ÷ 2 =	37)	60 ÷ 10 =
6)	42 ÷ 7 =	22)	108 ÷ 12 =	38)	8 ÷ 1 =
7)	54 ÷ 9 =	23)	33 ÷ 11 =	39)	144 ÷ 12 =
8)	52 ÷ 13 =	24)	20 ÷ 10 =	40)	35 ÷ 7 =
9)	15 ÷ 3 =	25)	36 ÷ 4 =	41)	55 ÷ 11=
10)	35 ÷ 5 =	26)	30 ÷ 6 =	42)	63 ÷ 7 =
11)	20 ÷ 2 =	27)	64 ÷ 8 =	43)	169 ÷ 13 =
12)	12 ÷ 6 =	28)	96 ÷ 12 =	44)	4 ÷ 2 =
13)	18 ÷ 9 =	29)	77 ÷ 11 =	45)	25 ÷ 5 =
14)	0 ÷ 1 =	30)	21 ÷ 3 =	46)	16 ÷ 8 =
15)	63 ÷ 7 =	31)	100 ÷ 10 =	47)	18 ÷ 3 =
16)	18 ÷ 2 =	32)	6 ÷ 2 =	48)	32 ÷ 4 =

DAY: 76

NAME: ____________________ TIME: ____ : ____ SCORE: ________

DIVIDE 0 - 13

1)	24 ÷ 3 =	17)	84 ÷ 12 =	33)	65 ÷ 13 =
2)	26 ÷ 13 =	18)	44 ÷ 11 =	34)	0 ÷ 6 =
3)	45 ÷ 9 =	19)	81 ÷ 9 =	35)	54 ÷ 9 =
4)	36 ÷ 12 =	20)	70 ÷ 10 =	36)	121 ÷ 11 =
5)	60 ÷ 10 =	21)	28 ÷ 4 =	37)	132 ÷ 12 =
6)	48 ÷ 6 =	22)	42 ÷ 7 =	38)	108 ÷ 12 =
7)	16 ÷ 2 =	23)	144 ÷ 12 =	39)	33 ÷ 11 =
8)	72 ÷ 8 =	24)	78 ÷ 13 =	40)	20 ÷ 10 =
9)	55 ÷ 11 =	25)	27 ÷ 3 =	41)	36 ÷ 4 =
10)	63 ÷ 7 =	26)	35 ÷ 5 =	42)	54 ÷ 6 =
11)	13 ÷ 13 =	27)	20 ÷ 2 =	43)	64 ÷ 8 =
12)	4 ÷ 2 =	28)	12 ÷ 6 =	44)	96 ÷ 12 =
13)	40 ÷ 5 =	29)	18 ÷ 9 =	45)	0 ÷ 10 =
14)	16 ÷ 8 =	30)	6 ÷ 1 =	46)	21 ÷ 3 =
15)	18 ÷ 3 =	31)	49 ÷ 7 =	47)	81 ÷ 9 =
16)	20 ÷ 4 =	32)	18 ÷ 2 =	48)	6 ÷ 2 =

ANSWER KEYS

DAY 1

1)	0	2)	8	3)	0	4)	5	5)	0	6)	7	7)	0	8)	4
9)	0	10)	3	11)	0	12)	9	13)	0	14)	0	15)	4	16)	3
17)	0	18)	9	19)	0	20)	4	21)	2	22)	0	23)	0	24)	6
25)	0	26)	1	27)	0	28)	7	29)	0	30)	1	31)	0	32)	2
33)	0	34)	4	35)	0	36)	3	37)	0	38)	3	39)	0	40)	0
41)	4	42)	1	43)	0	44)	5	45)	0	46)	6	47)	0	48)	8

DAY 2

1)	0	2)	9	3)	4	4)	7	5)	0	6)	3	7)	0	8)	4
9)	3	10)	0	11)	0	12)	3	13)	0	14)	4	15)	9	16)	0
17)	0	18)	3	19)	0	20)	0	21)	0	22)	2	23)	0	24)	6
25)	0	26)	1	27)	0	28)	7	29)	0	30)	1	31)	0	32)	2
33)	1	34)	8	35)	0	36)	5	37)	0	38)	7	39)	0	40)	4
41)	0	42)	1	43)	0	44)	5	45)	0	46)	4	47)	9	48)	0

DAY 3

1)	5	2)	0	3)	7	4)	3	5)	0	6)	8	7)	0	8)	0
9)	4	10)	0	11)	0	12)	4	13)	0	14)	3	15)	0	16)	9
17)	6	18)	0	19)	2	20)	8	21)	0	22)	9	23)	0	24)	0
25)	1	26)	0	27)	0	28)	6	29)	0	30)	1	31)	0	32)	7
33)	3	34)	0	35)	3	36)	9	37)	0	38)	4	39)	0	40)	0
41)	4	42)	0	43)	0	44)	4	45)	0	46)	0	47)	0	48)	5

DAY 4

1)	3	2)	0	3)	9	4)	0	5)	4	6)	0	7)	3	8)	4
9)	0	10)	1	11)	0	12)	0	13)	5	14)	4	15)	0	16)	9
17)	0	18)	9	19)	0	20)	4	21)	0	22)	2	23)	0	24)	0
25)	8	26)	0	27)	5	28)	0	29)	7	30)	0	31)	4	32)	0
33)	0	34)	5	35)	0	36)	3	37)	0	38)	3	39)	0	40)	6
41)	0	42)	1	43)	0	44)	7	45)	1	46)	0	47)	0	48)	2

DAY 5

1)	3	2)	4	3)	5	4)	2	5)	1	6)	7	7)	3	8)	7
9)	5	10)	6	11)	0	12)	9	13)	8	14)	7	15)	2	16)	3
17)	7	18)	10	19)	2	20)	7	21)	4	22)	6	23)	1	24)	9
25)	5	26)	8	27)	4	28)	5	29)	3	30)	5	31)	3	32)	2
33)	5	34)	4	35)	5	36)	8	37)	3	38)	2	39)	6	40)	4
41)	3	42)	1	43)	2	44)	6	45)	2	46)	10	47)	9	48)	4

DAY 6

1)	7	2)	10	3)	2	4)	8	5)	4	6)	6	7)	9	8)	4
9)	3	10)	1	11)	2	12)	9	13)	8	14)	7	15)	2	16)	3
17)	3	18)	4	19)	5	20)	2	21)	1	22)	7	23)	3	24)	2
25)	5	26)	8	27)	4	28)	6	29)	2	30)	10	31)	9	32)	4
33)	1	34)	4	35)	5	36)	8	37)	3	38)	2	39)	0	40)	7
41)	5	42)	6	43)	2	44)	5	45)	3	46)	5	47)	1	48)	2

DAY 7

1)	4	2)	7	3)	5	4)	6	5)	2	6)	5	7)	6	8)	4
9)	3	10)	6	11)	7	12)	10	13)	2	14)	7	15)	1	16)	3
17)	9	18)	4	19)	5	20)	2	21)	1	22)	7	23)	5	24)	4
25)	5	26)	8	27)	3	28)	10	29)	0	30)	4	31)	9	32)	10
33)	6	34)	2	35)	5	36)	8	37)	4	38)	1	39)	2	40)	9
41)	8	42)	7	43)	4	44)	6	45)	3	46)	5	47)	3	48)	2

DAY 8

1)	4	2)	7	3)	5	4)	6	5)	2	6)	6	7)	2	8)	5
9)	8	10)	4	11)	1	12)	10	13)	2	14)	7	15)	8	16)	3
17)	3	18)	4	19)	0	20)	2	21)	9	22)	8	23)	7	24)	4
25)	5	26)	8	27)	3	28)	2	29)	1	30)	10	31)	9	32)	4
33)	5	34)	6	35)	4	36)	3	37)	6	38)	7	39)	2	40)	1
41)	7	42)	5	43)	4	44)	6	45)	3	46)	5	47)	3	48)	2

DAY 9

1)	3	2)	5	3)	2	4)	6	5)	7	6)	0	7)	10	8)	3
9)	2	10)	1	11)	0	12)	8	13)	0	14)	9	15)	0	16)	6
17)	3	18)	0	19)	5	20)	0	21)	2	22)	5	23)	0	24)	9
25)	2	26)	4	27)	0	28)	4	29)	1	30)	3	31)	10	32)	0
33)	0	34)	6	35)	9	36)	0	37)	0	38)	8	39)	10	40)	7
41)	2	42)	6	43)	8	44)	0	45)	5	46)	1	47)	7	48)	4

DAY 10

1)	4	2)	6	3)	2	4)	0	5)	9	6)	4	7)	10	8)	0
9)	2	10)	0	11)	0	12)	5	13)	7	14)	9	15)	8	16)	3
17)	0	18)	6	19)	9	20)	0	21)	0	22)	8	23)	10	24)	7
25)	2	26)	6	27)	8	28)	0	29)	5	30)	1	31)	7	32)	4
33)	3	34)	0	35)	8	36)	0	37)	3	38)	5	39)	0	40)	9
41)	1	42)	5	43)	0	44)	4	45)	1	46)	3	47)	10	48)	0

DAY 11

1)	3	2)	2	3)	10	4)	9	5)	4	6)	5	7)	8	8)	2
9)	7	10)	8	11)	1	12)	9	13)	4	14)	6	15)	7	16)	1
17)	6	18)	3	19)	2	20)	4	21)	10	22)	7	23)	8	24)	9
25)	0	26)	7	27)	8	28)	9	29)	10	30)	2	31)	3	32)	7
33)	1	34)	6	35)	5	36)	4	37)	6	38)	3	39)	4	40)	2
41)	3	42)	4	43)	1	44)	5	45)	6	46)	10	47)	1	48)	6

DAY 12

1)	3	2)	2	3)	10	4)	9	5)	4	6)	5	7)	8	8)	9
9)	3	10)	4	11)	1	12)	5	13)	6	14)	10	15)	1	16)	6
17)	1	18)	6	19)	5	20)	4	21)	6	22)	0	23)	4	24)	2
25)	3	26)	7	27)	8	28)	9	29)	10	30)	2	31)	3	32)	7
33)	8	34)	3	35)	2	36)	4	37)	10	38)	7	39)	8	40)	2
41)	7	42)	8	43)	1	44)	9	45)	4	46)	6	47)	7	48)	1

DAY 13

1)	1	2)	9	3)	4	4)	6	5)	7	6)	0	7)	6	8)	5
9)	4	10)	6	11)	3	12)	4	13)	7	14)	10	15)	8	16)	6
17)	5	18)	8	19)	9	20)	3	21)	4	22)	2	23)	3	24)	1
25)	5	26)	6	27)	8	28)	9	29)	10	30)	2	31)	3	32)	7
33)	6	34)	3	35)	2	36)	10	37)	9	38)	7	39)	8	40)	2
41)	7	42)	8	43)	3	44)	2	45)	4	46)	10	47)	5	48)	1

DAY 14

1)	5	2)	3	3)	1	4)	10	5)	9	6)	7	7)	8	8)	2
9)	7	10)	8	11)	3	12)	2	13)	4	14)	10	15)	4	16)	0
17)	10	18)	9	19)	4	20)	6	21)	7	22)	1	23)	6	24)	5
25)	0	26)	6	27)	3	28)	4	29)	7	30)	5	31)	1	32)	6
33)	3	34)	8	35)	9	36)	5	37)	4	38)	2	39)	10	40)	1
41)	5	42)	6	43)	8	44)	9	45)	10	46)	2	47)	3	48)	7

DAY 15

1)	2	2)	4	3)	7	4)	9	5)	2	6)	6	7)	0	8)	7
9)	4	10)	6	11)	8	12)	0	13)	8	14)	7	15)	7	16)	0
17)	0	18)	1	19)	0	20)	7	21)	7	22)	0	23)	5	24)	3
25)	3	26)	10	27)	5	28)	9	29)	7	30)	4	31)	0	32)	2
33)	1	34)	5	35)	9	36)	5	37)	3	38)	6	39)	6	40)	0
41)	6	42)	1	43)	8	44)	6	45)	5	46)	10	47)	9	48)	0

DAY 16

1)	3	2)	10	3)	4	4)	9	5)	7	6)	4	7)	0	8)	2
9)	8	10)	5	11)	0	12)	7	13)	7	14)	0	15)	5	16)	3
17)	6	18)	9	19)	0	20)	6	21)	5	22)	10	23)	9	24)	1
25)	0	26)	4	27)	9	28)	5	29)	3	30)	6	31)	6	32)	0
33)	4	34)	1	35)	8	36)	0	37)	8	38)	7	39)	7	40)	0
41)	2	42)	4	43)	7	44)	9	45)	2	46)	6	47)	0	48)	1

DAY 17

1)	5	2)	6	3)	7	4)	4	5)	8	6)	1	7)	6	8)	7
9)	4	10)	3	11)	6	12)	9	13)	7	14)	3	15)	8	16)	3
17)	10	18)	4	19)	2	20)	3	21)	8	22)	2	23)	4	24)	3
25)	0	26)	2	27)	5	28)	2	29)	3	30)	9	31)	10	32)	5
33)	9	34)	1	35)	0	36)	4	37)	9	38)	10	39)	5	40)	2
41)	1	42)	10	43)	7	44)	4	45)	8	46)	6	47)	9	48)	1

DAY 18

1)	10	2)	3	3)	6	4)	9	5)	7	6)	3	7)	8	8)	5
9)	3	10)	6	11)	9	12)	2	13)	5	14)	2	15)	3	16)	10
17)	9	18)	1	19)	0	20)	4	21)	9	22)	10	23)	5	24)	6
25)	10	26)	4	27)	2	28)	6	29)	8	30)	2	31)	4	32)	7
33)	1	34)	10	35)	7	36)	4	37)	5	38)	6	39)	9	40)	1
41)	7	42)	4	43)	8	44)	1	45)	6	46)	7	47)	4	48)	2

DAY 19

1)	8	2)	6	3)	1	4)	10	5)	7	6)	5	7)	3	8)	6
9)	9	10)	1	11)	6	12)	9	13)	7	14)	3	15)	8	16)	1
17)	10	18)	4	19)	2	20)	3	21)	8	22)	7	23)	4	24)	8
25)	0	26)	6	27)	7	28)	4	29)	9	30)	10	31)	5	32)	7
33)	2	34)	4	35)	3	36)	9	37)	1	38)	3	39)	4	40)	2
41)	1	42)	10	43)	5	44)	2	45)	0	46)	9	47)	10	48)	6

DAY 20

1)	2	2)	5	3)	6	4)	7	5)	9	6)	1	7)	6	8)	4
9)	9	10)	10	11)	5	12)	3	13)	2	14)	4	15)	5	16)	1
17)	8	18)	3	19)	7	20)	5	21)	4	22)	8	23)	1	24)	2
25)	3	26)	9	27)	10	28)	5	29)	10	30)	4	31)	2	32)	3
33)	7	34)	0	35)	8	36)	2	37)	1	38)	10	39)	7	40)	4
41)	0	42)	6	43)	9	44)	1	45)	4	46)	3	47)	6	48)	9

DAY 21

1)	4	2)	6	3)	1	4)	9	5)	3	6)	2	7)	10	8)	0
9)	2	10)	6	11)	0	12)	8	13)	1	14)	9	15)	5	16)	3
17)	3	18)	0	19)	4	20)	1	21)	7	22)	5	23)	0	24)	6
25)	7	26)	10	27)	9	28)	2	29)	10	30)	7	31)	10	32)	0
33)	0	34)	3	35)	9	36)	0	37)	6	38)	8	39)	8	40)	7
41)	2	42)	9	43)	8	44)	0	45)	5	46)	1	47)	5	48)	4

DAY 22

1)	0	2)	10	3)	9	4)	0	5)	6	6)	8	7)	7	8)	8
9)	2	10)	9	11)	2	12)	0	13)	5	14)	1	15)	5	16)	4
17)	4	18)	6	19)	1	20)	1	21)	3	22)	2	23)	10	24)	0
25)	2	26)	6	27)	0	28)	8	29)	8	30)	9	31)	5	32)	3
33)	3	34)	0	35)	4	36)	3	37)	6	38)	5	39)	0	40)	6
41)	7	42)	3	43)	9	44)	7	45)	7	46)	7	47)	10	48)	0

DAY 23

1)	7	2)	2	3)	3	4)	8	5)	6	6)	4	7)	5	8)	9
9)	1	10)	3	11)	4	12)	2	13)	10	14)	8	15)	4	16)	9
17)	8	18)	3	19)	10	20)	2	21)	6	22)	7	23)	4	24)	7
25)	3	26)	5	27)	8	28)	0	29)	2	30)	6	31)	8	32)	4
33)	3	34)	2	35)	1	36)	6	37)	5	38)	9	39)	6	40)	1
41)	10	42)	8	43)	7	44)	10	45)	5	46)	1	47)	9	48)	8

DAY 24

1)	3	2)	2	3)	1	4)	6	5)	5	6)	9	7)	5	8)	0
9)	1	10)	3	11)	4	12)	2	13)	10	14)	8	15)	4	16)	9
17)	8	18)	9	19)	10	20)	2	21)	0	22)	7	23)	4	24)	9
25)	10	26)	8	27)	7	28)	10	29)	2	30)	1	31)	10	32)	4
33)	7	34)	2	35)	3	36)	8	37)	6	38)	4	39)	6	40)	1
41)	3	42)	5	43)	8	44)	1	45)	7	46)	6	47)	9	48)	8

DAY 25

1)	3	2)	4	3)	7	4)	10	5)	5	6)	9	7)	5	8)	9
9)	1	10)	3	11)	2	12)	1	13)	6	14)	8	15)	4	16)	9
17)	9	18)	3	19)	10	20)	5	21)	8	22)	1	23)	2	24)	6
25)	9	26)	8	27)	7	28)	10	29)	2	30)	0	31)	8	32)	4
33)	7	34)	2	35)	3	36)	8	37)	6	38)	4	39)	7	40)	1
41)	3	42)	2	43)	6	44)	7	45)	5	46)	9	47)	1	48)	10

DAY 26

1)	6	2)	1	3)	3	4)	2	5)	6	6)	9	7)	1	8)	8
9)	3	10)	10	11)	5	12)	8	13)	6	14)	8	15)	4	16)	9
17)	9	18)	0	19)	3	20)	7	21)	1	22)	9	23)	2	24)	6
25)	9	26)	8	27)	7	28)	10	29)	2	30)	1	31)	5	32)	4
33)	7	34)	2	35)	3	36)	8	37)	6	38)	4	39)	3	40)	4
41)	2	42)	10	43)	5	44)	7	45)	4	46)	9	47)	10	48)	8

DAY 27

1)	5	2)	1	3)	7	4)	9	5)	3	6)	1	7)	9	8)	7
9)	4	10)	6	11)	5	12)	0	13)	10	14)	0	15)	7	16)	0
17)	1	18)	3	19)	8	20)	4	21)	0	22)	2	23)	8	24)	10
25)	6	26)	9	27)	1	28)	5	29)	3	30)	2	31)	10	32)	10
33)	5	34)	3	35)	6	36)	0	37)	9	38)	9	39)	0	40)	7
41)	6	42)	1	43)	8	44)	4	45)	5	46)	8	47)	8	48)	10

DAY 28

1)	1	2)	3	3)	8	4)	4	5)	0	6)	2	7)	8	8)	3
9)	6	10)	9	11)	1	12)	5	13)	1	14)	2	15)	10	16)	6
17)	5	18)	3	19)	9	20)	0	21)	9	22)	9	23)	0	24)	7
25)	6	26)	10	27)	6	28)	4	29)	5	30)	8	31)	8	32)	10
33)	5	34)	8	35)	7	36)	9	37)	3	38)	2	39)	9	40)	7
41)	4	42)	6	43)	5	44)	0	45)	10	46)	0	47)	7	48)	0

DAY 29

1)	6	2)	1	3)	3	4)	2	5)	5	6)	4	7)	9	8)	7
9)	8	10)	1	11)	2	12)	3	13)	10	14)	6	15)	8	16)	10
17)	7	18)	3	19)	10	20)	0	21)	5	22)	6	23)	8	24)	7
25)	9	26)	2	27)	3	28)	1	29)	9	30)	7	31)	4	32)	6
33)	5	34)	2	35)	9	36)	10	37)	4	38)	6	39)	5	40)	2
41)	10	42)	8	43)	6	44)	4	45)	0	46)	1	47)	5	48)	7

DAY 30

1)	5	2)	2	3)	9	4)	10	5)	6	6)	1	7)	3	8)	2
9)	5	10)	4	11)	3	12)	0	13)	10	14)	6	15)	8	16)	4
17)	7	18)	3	19)	8	20)	7	21)	9	22)	0	23)	5	24)	2
25)	10	26)	8	27)	2	28)	1	29)	9	30)	7	31)	8	32)	6
33)	4	34)	6	35)	10	36)	1	37)	5	38)	6	39)	9	40)	7
41)	8	42)	4	43)	6	44)	10	45)	3	46)	1	47)	5	48)	7

DAY 31

1)	5	2)	4	3)	9	4)	10	5)	4	6)	0	7)	8	8)	7
9)	9	10)	2	11)	3	12)	4	13)	10	14)	6	15)	8	16)	10
17)	7	18)	3	19)	8	20)	1	21)	5	22)	6	23)	9	24)	0
25)	8	26)	1	27)	2	28)	1	29)	9	30)	7	31)	0	32)	6
33)	6	34)	1	35)	3	36)	2	37)	5	38)	4	39)	5	40)	2
41)	10	42)	8	43)	6	44)	10	45)	3	46)	1	47)	5	48)	7

DAY 32

1)	7	2)	8	3)	1	4)	6	5)	2	6)	1	7)	3	8)	2
9)	5	10)	4	11)	1	12)	3	13)	10	14)	6	15)	8	16)	10
17)	7	18)	3	19)	8	20)	7	21)	9	22)	2	23)	5	24)	0
25)	10	26)	8	27)	2	28)	4	29)	9	30)	7	31)	8	32)	6
33)	4	34)	6	35)	10	36)	1	37)	5	38)	6	39)	9	40)	5
41)	2	42)	9	43)	10	44)	1	45)	2	46)	6	47)	5	48)	7

DAY 33

1)	4	2)	9	3)	7	4)	7	5)	2	6)	0	7)	3	8)	6
9)	4	10)	5	11)	3	12)	0	13)	8	14)	1	15)	5	16)	0
17)	2	18)	8	19)	8	20)	9	21)	0	22)	9	23)	9	24)	7
25)	5	26)	3	27)	4	28)	5	29)	3	30)	4	31)	9	32)	10
33)	2	34)	7	35)	5	36)	6	37)	6	38)	9	39)	0	40)	1
41)	7	42)	6	43)	2	44)	3	45)	7	46)	8	47)	10	48)	8

DAY 34

1)	2	2)	8	3)	8	4)	9	5)	0	6)	9	7)	9	8)	7
9)	5	10)	3	11)	4	12)	5	13)	3	14)	4	15)	9	16)	10
17)	2	18)	6	19)	5	20)	6	21)	6	22)	9	23)	0	24)	1
25)	7	26)	6	27)	2	28)	3	29)	7	30)	8	31)	10	32)	8
33)	4	34)	9	35)	7	36)	7	37)	2	38)	0	39)	6	40)	4
41)	4	42)	5	43)	3	44)	0	45)	10	46)	1	47)	5	48)	0

DAY 35

1)	3	2)	1	3)	0	4)	4	5)	5	6)	7	7)	3	8)	6
9)	8	10)	4	11)	10	12)	9	13)	3	14)	5	15)	7	16)	1
17)	6	18)	2	19)	7	20)	1	21)	8	22)	4	23)	8	24)	9
25)	10	26)	0	27)	2	28)	4	29)	5	30)	3	31)	9	32)	2
33)	5	34)	4	35)	8	36)	10	37)	9	38)	6	39)	2	40)	3
41)	10	42)	7	43)	6	44)	1	45)	2	46)	10	47)	5	48)	6

DAY 36

1)	6	2)	2	3)	7	4)	8	5)	4	6)	2	7)	4	8)	9
9)	10	10)	1	11)	10	12)	9	13)	3	14)	5	15)	7	16)	1
17)	3	18)	1	19)	2	20)	4	21)	5	22)	7	23)	0	24)	7
25)	1	26)	8	27)	4	28)	8	29)	5	30)	3	31)	9	32)	2
33)	5	34)	4	35)	8	36)	10	37)	9	38)	6	39)	2	40)	3
41)	10	42)	7	43)	6	44)	1	45)	0	46)	10	47)	5	48)	6

DAY 37

1)	8	2)	10	3)	9	4)	5	5)	2	6)	8	7)	4	8)	3
9)	10	10)	1	11)	10	12)	9	13)	3	14)	5	15)	7	16)	1
17)	3	18)	1	19)	2	20)	4	21)	5	22)	7	23)	2	24)	7
25)	1	26)	8	27)	4	28)	8	29)	5	30)	3	31)	7	32)	9
33)	5	34)	4	35)	6	36)	8	37)	6	38)	2	39)	4	40)	9
41)	10	42)	7	43)	6	44)	1	45)	2	46)	10	47)	0	48)	6

DAY 38

1)	1	2)	2	3)	10	4)	1	5)	9	6)	3	7)	6	8)	7
9)	1	10)	8	11)	10	12)	9	13)	5	14)	10	15)	7	16)	6
17)	7	18)	0	19)	8	20)	4	21)	8	22)	5	23)	3	24)	7
25)	2	26)	3	27)	1	28)	2	29)	4	30)	5	31)	7	32)	2
33)	9	34)	2	35)	8	36)	4	37)	3	38)	10	39)	1	40)	5
41)	6	42)	5	43)	4	44)	6	45)	2	46)	6	47)	0	48)	4

DAY 39

1)	6	2)	3	3)	9	4)	9	5)	9	6)	0	7)	3	8)	5
9)	10	10)	8	11)	9	12)	0	13)	1	14)	10	15)	2	16)	4
17)	2	18)	4	19)	6	20)	3	21)	0	22)	9	23)	10	24)	5
25)	10	26)	3	27)	1	28)	5	29)	6	30)	2	31)	3	32)	5
33)	3	34)	2	35)	7	36)	10	37)	7	38)	10	39)	7	40)	6
41)	2	42)	7	43)	4	44)	8	45)	10	46)	1	47)	8	48)	1

DAY 40

1)	3	2)	2	3)	5	4)	9	5)	7	6)	10	7)	6	8)	3
9)	2	10)	7	11)	9	12)	8	13)	10	14)	4	15)	8	16)	1
17)	6	18)	3	19)	5	20)	9	21)	6	22)	0	23)	3	24)	5
25)	4	26)	8	27)	7	28)	0	29)	1	30)	10	31)	2	32)	7
33)	2	34)	4	35)	6	36)	3	37)	0	38)	9	39)	10	40)	8
41)	10	42)	6	43)	0	44)	5	45)	4	46)	2	47)	9	48)	5

DAY 41

1)	2	2)	5	3)	4	4)	7	5)	1	6)	10	7)	9	8)	8
9)	5	10)	7	11)	10	12)	8	13)	3	14)	4	15)	6	16)	1
17)	6	18)	8	19)	5	20)	1	21)	7	22)	2	23)	3	24)	8
25)	10	26)	0	27)	4	28)	3	29)	1	30)	9	31)	6	32)	7
33)	0	34)	4	35)	2	36)	3	37)	5	38)	9	39)	8	40)	10
41)	6	42)	3	43)	5	44)	7	45)	9	46)	10	47)	2	48)	9

DAY 42

1)	3	2)	5	3)	9	4)	8	5)	10	6)	0	7)	3	8)	8
9)	5	10)	7	11)	10	12)	8	13)	3	14)	4	15)	0	16)	1
17)	1	18)	4	19)	2	20)	1	21)	7	22)	2	23)	3	24)	8
25)	10	26)	6	27)	4	28)	3	29)	1	30)	9	31)	6	32)	7
33)	6	34)	8	35)	5	36)	2	37)	5	38)	4	39)	7	40)	1
41)	10	42)	9	43)	5	44)	7	45)	9	46)	10	47)	2	48)	9

DAY 43

1)	3	2)	8	3)	5	4)	2	5)	6	6)	4	7)	2	8)	1
9)	7	10)	2	11)	3	12)	8	13)	3	14)	4	15)	2	16)	6
17)	1	18)	4	19)	3	20)	8	21)	5	22)	7	23)	10	24)	8
25)	0	26)	2	27)	4	28)	3	29)	1	30)	9	31)	6	32)	7
33)	6	34)	5	35)	9	36)	8	37)	10	38)	2	39)	7	40)	1
41)	10	42)	9	43)	5	44)	7	45)	9	46)	10	47)	0	48)	9

DAY 44

1)	2	2)	5	3)	4	4)	7	5)	1	6)	10	7)	9	8)	1
9)	4	10)	2	11)	3	12)	8	13)	3	14)	4	15)	2	16)	1
17)	6	18)	8	19)	10	20)	0	21)	3	22)	5	23)	3	24)	8
25)	10	26)	2	27)	4	28)	3	29)	0	30)	9	31)	6	32)	7
33)	8	34)	5	35)	7	36)	10	37)	5	38)	9	39)	8	40)	5
41)	1	42)	7	43)	2	44)	6	45)	9	46)	10	47)	2	48)	9

DAY 45

1)	9	2)	8	3)	7	4)	5	5)	6	6)	0	7)	3	8)	10
9)	3	10)	8	11)	6	12)	1	13)	9	14)	8	15)	6	16)	7
17)	3	18)	3	19)	9	20)	7	21)	0	22)	8	23)	5	24)	1
25)	10	26)	4	27)	4	28)	5	29)	5	30)	2	31)	8	32)	8
33)	2	34)	9	35)	5	36)	9	37)	9	38)	10	39)	9	40)	6
41)	9	42)	3	43)	3	44)	4	45)	10	46)	7	47)	6	48)	2

DAY 46

1)	3	2)	3	3)	9	4)	7	5)	0	6)	8	7)	5	8)	1
9)	10	10)	6	11)	4	12)	5	13)	7	14)	3	15)	8	16)	6
17)	5	18)	9	19)	5	20)	9	21)	10	22)	9	23)	9	24)	6
25)	3	26)	3	27)	3	28)	4	29)	10	30)	7	31)	6	32)	2
33)	7	34)	8	35)	7	36)	7	37)	2	38)	0	39)	4	40)	10
41)	3	42)	8	43)	4	44)	1	45)	9	46)	8	47)	6	48)	8

DAY 47

1)	2	2)	5	3)	4	4)	1	5)	9	6)	10	7)	8	8)	7
9)	5	10)	6	11)	10	12)	8	13)	9	14)	4	15)	2	16)	3
17)	6	18)	7	19)	3	20)	1	21)	6	22)	2	23)	1	24)	7
25)	4	26)	9	27)	4	28)	3	29)	10	30)	8	31)	5	32)	6
33)	1	34)	4	35)	2	36)	3	37)	9	38)	8	39)	7	40)	10
41)	3	42)	2	43)	5	44)	6	45)	0	46)	10	47)	9	48)	8

DAY 48

1)	2	2)	3	3)	9	4)	8	5)	7	6)	10	7)	8	8)	7
9)	5	10)	6	11)	10	12)	8	13)	9	14)	4	15)	2	16)	3
17)	6	18)	7	19)	10	20)	3	21)	2	22)	5	23)	6	24)	7
25)	4	26)	9	27)	4	28)	3	29)	10	30)	8	31)	5	32)	6
33)	1	34)	4	35)	2	36)	5	37)	4	38)	1	39)	0	40)	3
41)	1	42)	6	43)	2	44)	1	45)	10	46)	7	47)	9	48)	8

DAY 49

1)	4	2)	3	3)	10	4)	8	5)	2	6)	10	7)	8	8)	7
9)	5	10)	6	11)	10	12)	8	13)	9	14)	4	15)	0	16)	3
17)	6	18)	7	19)	10	20)	3	21)	2	22)	5	23)	6	24)	7
25)	1	26)	4	27)	2	28)	5	29)	4	30)	1	31)	9	32)	3
33)	4	34)	9	35)	2	36)	3	37)	9	38)	8	39)	7	40)	6
41)	1	42)	5	43)	2	44)	7	45)	1	46)	10	47)	9	48)	8

DAY 50

1)	6	2)	0	3)	8	4)	7	5)	2	6)	10	7)	8	8)	7
9)	5	10)	6	11)	10	12)	8	13)	9	14)	4	15)	2	16)	3
17)	3	18)	9	19)	2	20)	4	21)	1	22)	3	23)	4	24)	6
25)	10	26)	6	27)	2	28)	9	29)	3	30)	8	31)	9	32)	7
33)	5	34)	7	35)	10	36)	3	37)	2	38)	5	39)	6	40)	7
41)	1	42)	4	43)	2	44)	5	45)	4	46)	1	47)	9	48)	3

DAY 51

1)	9	2)	6	3)	7	4)	3	5)	3	6)	0	7)	1	8)	4
9)	6	10)	4	11)	0	12)	9	13)	4	14)	9	15)	7	16)	7
17)	6	18)	5	19)	6	20)	3	21)	0	22)	8	23)	8	24)	4
25)	5	26)	9	27)	0	28)	6	29)	4	30)	7	31)	2	32)	8
33)	2	34)	8	35)	7	36)	4	37)	10	38)	2	39)	2	40)	7
41)	5	42)	8	43)	8	44)	5	45)	10	46)	4	47)	9	48)	7

DAY 52

1)	3	2)	8	3)	7	4)	4	5)	10	6)	2	7)	2	8)	9
9)	7	10)	8	11)	8	12)	4	13)	10	14)	4	15)	4	16)	7
17)	6	18)	5	19)	6	20)	3	21)	0	22)	8	23)	8	24)	4
25)	5	26)	9	27)	9	28)	4	29)	6	30)	7	31)	2	32)	8
33)	0	34)	6	35)	7	36)	5	37)	3	38)	0	39)	1	40)	4
41)	5	42)	3	43)	0	44)	7	45)	5	46)	9	47)	7	48)	8

DAY 53

1)	8	2)	4	3)	3	4)	1	5)	8	6)	10	7)	7	8)	6
9)	4	10)	1	11)	10	12)	7	13)	9	14)	3	15)	1	16)	2
17)	5	18)	6	19)	2	20)	0	21)	5	22)	1	23)	10	24)	6
25)	3	26)	0	27)	3	28)	2	29)	10	30)	7	31)	6	32)	0
33)	9	34)	8	35)	5	36)	7	37)	9	38)	3	39)	4	40)	10
41)	5	42)	2	43)	4	44)	5	45)	1	46)	10	47)	8	48)	7

DAY 54

1)	5	2)	1	3)	4	4)	6	5)	9	6)	4	7)	3	8)	1
9)	4	10)	5	11)	10	12)	7	13)	8	14)	3	15)	1	16)	2
17)	6	18)	4	19)	9	20)	8	21)	1	22)	7	23)	6	24)	2
25)	1	26)	5	27)	3	28)	2	29)	10	30)	7	31)	4	32)	5
33)	0	34)	10	35)	6	36)	3	37)	8	38)	5	39)	7	40)	2
41)	7	42)	10	43)	6	44)	5	45)	4	46)	9	47)	8	48)	7

DAY 55

1)	2	2)	10	3)	7	4)	6	5)	8	6)	4	7)	3	8)	1
9)	4	10)	9	11)	10	12)	7	13)	8	14)	3	15)	1	16)	2
17)	0	18)	8	19)	5	20)	7	21)	8	22)	1	23)	10	24)	6
25)	3	26)	8	27)	3	28)	2	29)	10	30)	7	31)	4	32)	5
33)	9	34)	6	35)	2	36)	0	37)	5	38)	3	39)	4	40)	10
41)	5	42)	2	43)	4	44)	5	45)	1	46)	9	47)	8	48)	6

DAY 56

1)	5	2)	1	3)	4	4)	6	5)	8	6)	4	7)	3	8)	1
9)	4	10)	5	11)	9	12)	7	13)	8	14)	3	15)	1	16)	2
17)	0	18)	8	19)	2	20)	7	21)	2	22)	9	23)	6	24)	2
25)	1	26)	5	27)	3	28)	2	29)	10	30)	7	31)	4	32)	5
33)	1	34)	10	35)	6	36)	3	37)	8	38)	6	39)	4	40)	10
41)	8	42)	10	43)	7	44)	5	45)	0	46)	9	47)	6	48)	7

DAY 57

1)	8	2)	7	3)	9	4)	1	5)	5	6)	9	7)	7	8)	4
9)	3	10)	8	11)	0	12)	8	13)	2	14)	10	15)	6	16)	9
17)	5	18)	10	19)	9	20)	7	21)	1	22)	4	23)	4	24)	2
25)	8	26)	4	27)	6	28)	5	29)	4	30)	3	31)	6	32)	3
33)	5	34)	10	35)	7	36)	7	37)	2	38)	8	39)	1	40)	0
41)	9	42)	8	43)	8	44)	6	45)	5	46)	9	47)	6	48)	9

DAY 58

1)	5	2)	10	3)	7	4)	8	5)	2	6)	9	7)	1	8)	6
9)	1	10)	8	11)	9	12)	6	13)	5	14)	9	15)	6	16)	9
17)	5	18)	6	19)	9	20)	7	21)	1	22)	6	23)	4	24)	2
25)	8	26)	4	27)	10	28)	5	29)	4	30)	3	31)	9	32)	3
33)	8	34)	7	35)	4	36)	1	37)	5	38)	9	39)	7	40)	4
41)	3	42)	9	43)	0	44)	8	45)	2	46)	8	47)	6	48)	8

DAY 59

1)	8	2)	4	3)	3	4)	6	5)	0	6)	11	7)	7	8)	6
9)	4	10)	5	11)	10	12)	7	13)	9	14)	3	15)	1	16)	2
17)	5	18)	6	19)	2	20)	1	21)	5	22)	2	23)	10	24)	6
25)	3	26)	8	27)	3	28)	2	29)	10	30)	7	31)	4	32)	5
33)	9	34)	8	35)	0	36)	7	37)	9	38)	3	39)	4	40)	10
41)	5	42)	0	43)	4	44)	5	45)	1	46)	11	47)	8	48)	7

DAY 60

1)	3	2)	4	3)	11	4)	5	5)	1	6)	10	7)	7	8)	5
9)	6	10)	2	11)	1	12)	7	13)	8	14)	3	15)	1	16)	2
17)	6	18)	4	19)	5	20)	9	21)	0	22)	2	23)	10	24)	6
25)	3	26)	9	27)	3	28)	2	29)	10	30)	7	31)	4	32)	6
33)	9	34)	8	35)	5	36)	7	37)	6	38)	8	39)	4	40)	3
41)	1	42)	8	43)	4	44)	5	45)	0	46)	11	47)	9	48)	7

DAY 61

1)	3	2)	2	3)	1	4)	7	5)	8	6)	10	7)	7	8)	5
9)	6	10)	2	11)	11	12)	5	13)	1	14)	3	15)	4	16)	2
17)	6	18)	4	19)	5	20)	11	21)	6	22)	1	23)	10	24)	6
25)	3	26)	0	27)	4	28)	2	29)	10	30)	7	31)	4	32)	5
33)	9	34)	8	35)	4	36)	5	37)	1	38)	0	39)	4	40)	3
41)	1	42)	8	43)	9	44)	7	45)	8	46)	0	47)	9	48)	7

DAY 62

1)	3	2)	10	3)	6	4)	3	5)	8	6)	5	7)	7	8)	9
9)	6	10)	2	11)	10	12)	5	13)	1	14)	3	15)	11	16)	2
17)	9	18)	11	19)	4	20)	5	21)	1	22)	9	23)	4	24)	3
25)	1	26)	8	27)	3	28)	2	29)	0	30)	7	31)	4	32)	5
33)	6	34)	4	35)	5	36)	0	37)	8	38)	1	39)	2	40)	3
41)	7	42)	6	43)	11	44)	7	45)	8	46)	10	47)	9	48)	7

DAY 63

1)	9	2)	8	3)	6	4)	8	5)	5	6)	6	7)	8	8)	9
9)	6	10)	4	11)	1	12)	3	13)	4	14)	8	15)	7	16)	2
17)	5	18)	4	19)	7	20)	3	21)	2	22)	6	23)	5	24)	4
25)	4	26)	9	27)	9	28)	5	29)	3	30)	7	31)	9	32)	8
33)	3	34)	9	35)	5	36)	0	37)	10	38)	3	39)	2	40)	7
41)	6	42)	7	43)	0	44)	11	45)	6	46)	3	47)	10	48)	5

DAY 64

1)	3	2)	9	3)	5	4)	4	5)	10	6)	3	7)	2	8)	7
9)	6	10)	7	11)	9	12)	2	13)	6	14)	3	15)	10	16)	5
17)	7	18)	8	19)	6	20)	8	21)	6	22)	6	23)	8	24)	5
25)	6	26)	0	27)	1	28)	3	29)	4	30)	8	31)	7	32)	2
33)	5	34)	4	35)	0	36)	3	37)	2	38)	4	39)	5	40)	4
41)	4	42)	9	43)	9	44)	5	45)	3	46)	7	47)	9	48)	8

DAY 65

1)	8	2)	4	3)	3	4)	1	5)	8	6)	10	7)	7	8)	6
9)	12	10)	5	11)	10	12)	7	13)	8	14)	3	15)	9	16)	2
17)	12	18)	6	19)	2	20)	1	21)	5	22)	9	23)	11	24)	6
25)	0	26)	8	27)	3	28)	2	29)	10	30)	7	31)	4	32)	5
33)	9	34)	8	35)	0	36)	7	37)	8	38)	3	39)	4	40)	10
41)	5	42)	1	43)	4	44)	9	45)	1	46)	11	47)	3	48)	6

DAY 66

1)	12	2)	5	3)	1	4)	4	5)	11	6)	10	7)	7	8)	5
9)	6	10)	2	11)	1	12)	5	13)	8	14)	3	15)	9	16)	2
17)	6	18)	2	19)	5	20)	10	21)	7	22)	1	23)	10	24)	6
25)	3	26)	11	27)	9	28)	2	29)	10	30)	7	31)	4	32)	5
33)	9	34)	8	35)	5	36)	7	37)	0	38)	3	39)	4	40)	8
41)	4	42)	3	43)	1	44)	0	45)	12	46)	10	47)	8	48)	6

DAY 67

1)	10	2)	9	3)	8	4)	5	5)	7	6)	12	7)	8	8)	5
9)	6	10)	2	11)	1	12)	5	13)	8	14)	3	15)	1	16)	2
17)	6	18)	11	19)	4	20)	3	21)	1	22)	5	23)	10	24)	6
25)	3	26)	8	27)	0	28)	2	29)	10	30)	7	31)	4	32)	6
33)	5	34)	2	35)	4	36)	0	37)	9	38)	3	39)	4	40)	10
41)	11	42)	9	43)	7	44)	1	45)	6	46)	10	47)	12	48)	7

DAY 68

1)	10	2)	5	3)	8	4)	9	5)	12	6)	10	7)	7	8)	5
9)	6	10)	2	11)	1	12)	5	13)	8	14)	3	15)	1	16)	2
17)	6	18)	8	19)	4	20)	3	21)	1	22)	8	23)	0	24)	6
25)	3	26)	8	27)	11	28)	2	29)	10	30)	7	31)	4	32)	5
33)	1	34)	2	35)	4	36)	0	37)	11	38)	10	39)	4	40)	6
41)	5	42)	3	43)	7	44)	9	45)	0	46)	12	47)	6	48)	7

DAY 69

1)	2	2)	6	3)	8	4)	9	5)	4	6)	8	7)	7	8)	3
9)	8	10)	4	11)	3	12)	6	13)	6	14)	4	15)	6	16)	9
17)	5	18)	4	19)	9	20)	7	21)	2	22)	7	23)	8	24)	6
25)	4	26)	6	27)	0	28)	9	29)	2	30)	3	31)	5	32)	5
33)	5	34)	1	35)	7	36)	8	37)	2	38)	3	39)	3	40)	8
41)	6	42)	7	43)	10	44)	11	45)	6	46)	2	47)	1	48)	10

DAY 70

1)	5	2)	1	3)	7	4)	8	5)	9	6)	3	7)	3	8)	8
9)	6	10)	7	11)	10	12)	8	13)	6	14)	2	15)	1	16)	9
17)	5	18)	4	19)	9	20)	7	21)	2	22)	6	23)	8	24)	1
25)	4	26)	6	27)	7	28)	9	29)	2	30)	3	31)	11	32)	5
33)	2	34)	0	35)	3	36)	9	37)	5	38)	8	39)	9	40)	12
41)	8	42)	4	43)	2	44)	6	45)	9	46)	4	47)	6	48)	9

DAY 71

1)	8	2)	4	3)	3	4)	0	5)	8	6)	10	7)	7	8)	6
9)	4	10)	5	11)	10	12)	7	13)	8	14)	12	15)	1	16)	2
17)	5	18)	6	19)	2	20)	9	21)	5	22)	1	23)	10	24)	13
25)	3	26)	0	27)	3	28)	2	29)	11	30)	7	31)	4	32)	6
33)	9	34)	8	35)	5	36)	11	37)	3	38)	2	39)	12	40)	10
41)	5	42)	13	43)	4	44)	0	45)	1	46)	9	47)	8	48)	7

DAY 72

1)	7	2)	3	3)	4	4)	10	5)	3	6)	8	7)	9	8)	2
9)	0	10)	5	11)	10	12)	13	13)	6	14)	3	15)	1	16)	2
17)	0	18)	6	19)	2	20)	9	21)	5	22)	1	23)	10	24)	11
25)	8	26)	12	27)	7	28)	6	29)	10	30)	2	31)	4	32)	6
33)	11	34)	8	35)	5	36)	7	37)	8	38)	4	39)	3	40)	1
41)	13	42)	1	43)	4	44)	5	45)	1	46)	12	47)	0	48)	7

DAY 73

1)	6	2)	2	3)	4	4)	13	5)	10	6)	8	7)	3	8)	1
9)	4	10)	12	11)	3	12)	7	13)	8	14)	3	15)	1	16)	2
17)	5	18)	6	19)	2	20)	11	21)	4	22)	5	23)	1	24)	6
25)	0	26)	10	27)	7	28)	6	29)	11	30)	7	31)	4	32)	0
33)	9	34)	13	35)	5	36)	7	37)	2	38)	4	39)	3	40)	1
41)	5	42)	9	43)	5	44)	0	45)	12	46)	3	47)	13	48)	7

DAY 74

1)	7	2)	2	3)	12	4)	10	5)	7	6)	6	7)	3	8)	1
9)	4	10)	12	11)	3	12)	13	13)	11	14)	3	15)	1	16)	2
17)	9	18)	0	19)	5	20)	7	21)	8	22)	4	23)	1	24)	6
25)	8	26)	6	27)	10	28)	2	29)	9	30)	7	31)	4	32)	5
33)	11	34)	6	35)	2	36)	0	37)	4	38)	5	39)	3	40)	13
41)	5	42)	9	43)	0	44)	1	45)	10	46)	3	47)	8	48)	6

DAY 75

1)	7	2)	8	3)	9	4)	7	5)	7	6)	6	7)	6	8)	4
9)	5	10)	7	11)	10	12)	2	13)	2	14)	0	15)	9	16)	9
17)	5	18)	8	19)	0	20)	11	21)	7	22)	9	23)	3	24)	2
25)	9	26)	5	27)	8	28)	8	29)	7	30)	7	31)	10	32)	3
33)	8	34)	2	35)	5	36)	3	37)	6	38)	8	39)	12	40)	5
41)	5	42)	9	43)	13	44)	2	45)	5	46)	2	47)	6	48)	8

DAY 76

1)	8	2)	2	3)	5	4)	3	5)	6	6)	8	7)	8	8)	9
9)	5	10)	9	11)	1	12)	2	13)	8	14)	2	15)	6	16)	5
17)	7	18)	4	19)	9	20)	7	21)	7	22)	6	23)	12	24)	6
25)	9	26)	7	27)	10	28)	2	29)	2	30)	6	31)	7	32)	9
33)	5	34)	0	35)	6	36)	11	37)	11	38)	9	39)	3	40)	2
41)	9	42)	9	43)	8	44)	8	45)	0	46)	7	47)	9	48)	3

ABOUT THE AUTHOR

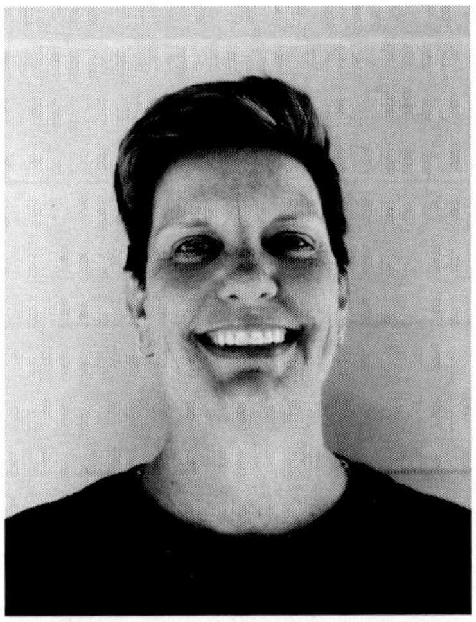

Karolyn Morris ("Karol") holds a Bachelor of Arts in Mathematics and a Master of Arts in Teaching Mathematics degrees and has been an educator for over 25 years. Her experience includes high school, K-12 charter school, and college math instruction.

Karol has a wide range of K-12 teaching experience, including high school, independent study, and assisting parents with homeschooling their K-12 children. She has held full-time faculty positions at Maui Community College (as a Developmental Mathematics Instructor) and North Georgia Technical College, teaching both learning support and college-level math courses. She has also taught online as Adjunct Math Faculty for a variety of colleges and universities.

Karol creates math videos, workbooks, and online courses to help students learn mathematics globally. Join Karol's newsletter to hear about new and upcoming releases, as well as participate in free book giveaways. Send an email titled "Free Stuff" to

karol@lastminutemathtutoring.com

JOIN MY NEWSLETTER

Get your FREE copy of Multiplication Math Drills (Tests with Answers)

https://BookHip.com/FJNQQH

Daily timed math drills provide the repetition needed to memorize math facts and develop math fluency. This book contains 16 days of timed multiplication worksheets (covering 0-13 multiplication facts). All answer keys are included.

NOTES

Made in United States
Cleveland, OH
06 November 2026

47662563R00057